D1112709

Praise for Plan B

"Lester Brown tells us how to build a more just world and save the planet...in a practical, straightforward way. We should all heed his advice."

— President Bill Clinton

"...a far-reaching thinker."

— *U.S. News & World Report*

"The best book on the environment I've ever read."

— Chris Swan, *Financial Times*

"It's exciting...a masterpiece!"

—Ted Turner

"[Brown's] ability to make a complicated subject accessible to the general reader is remarkable..."

— Katherine Salant, *Washington Post*

"In tackling a host of pressing issues in a single book, *Plan B 2.0* makes for an eye-opening read."

—*Times Higher Education Supplement*

"A great blueprint for combating climate change."

— Bryan Walsh, *Time*

"[Brown] lays out one of the most comprehensive set of solutions you can find in one place."

— Joseph Romm, *Climate Progress*

"...a highly readable and authoritative account of the problems we face from global warming to shrinking water resources, fisheries, forests, etc. The picture is very frightening. But the book also provides a way forward."

— Clare Short, British Member of Parliament

"Lester R. Brown gives concise, but very informative, summaries of what he regards as the key issues facing civilization as a consequence of the stress we put on our environment....a valuable contribution to the ongoing debate."

—*The Ecologist*

"Brown is impassioned and convincing when talking about the world's ills and what he considers the four great goals to restoring civilization's equilibrium..."

— April Streeter, *TreeHugger.com*

"In this impressively researched manifesto for change, Brown bluntly sets out the challenges and offers an achievable road map for solving the climate change crisis."

— *The Guardian*

"...the best summation of humanity's converging ecological problems and the best roadmap to solving them, all in one compact package."

— David Roberts, *Grist*

"Lester R. Brown...offers an attractive 21st-century alternative to the unacceptable business-as-usual path that we have been following with regard to the environment (Plan A), which is leading us to 'economic decline and collapse.'"

— Thomas F. Malone, *American Scientist*

"Brown's overall action plan is both comprehensive and compelling."

— Caroline Lucas, *Resurgence*

"Beautifully written and unimpeachably well-informed."

— Ross Gelbspan, author of *The Heat Is On*

"The best single volume on saving the earth, period."

— Geoffrey Holland, author of *The Hydrogen Age*

Full Planet, Empty Plates

OTHER NORTON BOOKS
BY LESTER R. BROWN

Earth Policy Institute® is a nonprofit environmental research organization providing a plan for building a sustainable future. In addition to the Plan B series, the Institute issues four-page *Plan B Updates* that assess progress in implementing Plan B. All of these plus additional data and graphs can be downloaded at no charge from www.earth-policy.org.

FULL PLANET, EMPTY PLATES

The New Geopolitics of Food Scarcity

Lester R. Brown

EARTH POLICY INSTITUTE

W • W • NORTON & COMPANY

NEW YORK   LONDON

The Earth Policy Institute trademark is registered in the U.S. Patent and
Trademark Office.

The views expressed are those of the author and do not necessarily represent those of the Earth Policy Institute; of its directors, officers, or staff;
or of any funders.

The text of this book is composed in Sabon. Composition by Elizabeth
Doherty; manufacturing by Maple Vale.

ISBN: 978-0-393-08891-5 (cloth) 978-0-393-34415-8 (pbk)

W. W. Norton & Company, Inc., 500 Fifth Avenue,
New York, N.Y. 10110
www.wwnorton.com

W. W. Norton & Company, Ltd., Castle House, 75/76 Wells Street,
London W1T 3QT

1 2 3 4 5 6 7 8 9 0

⊕ *This book is printed on recycled paper.*

In Memory of
Blondeen Gravely
1944–2012

Contents

Permission for reprinting or excerpting portions of the manuscript can be obtained from Reah Janise Kauffman at Earth Policy Institute. For full citations, data, and additional information on the topics discussed in this book, see www.earth-policy.org.

Preface

We started this book in the spring of 2012, corn planting time. U.S. farmers were planting some 96 million acres in corn, the most in 75 years. A warm early spring got the crop off to a great start. Analysts were predicting the largest corn harvest on record.

The United States is the world's leading producer and exporter of corn. At home, corn accounts for four fifths of the U.S. grain harvest. Internationally, the U.S. corn crop exceeds China's rice and wheat harvests combined. While wheat and rice are the world's leading food grains, corn totally dominates the use of grain in livestock and poultry feed.

The U.S. corn crop is as sensitive as it is productive. A thirsty, fast-growing plant, corn is vulnerable to both extreme heat and drought. At elevated temperatures, the corn plant, which is normally so productive, goes into thermal shock.

As spring turned into summer, the thermometer began to rise across the Corn Belt. In St. Louis, Missouri, in the southern Corn Belt, the temperature in late June and early July climbed to 100 degrees or higher 10 days in a row. The entire Corn Belt was blanketed with dehydrating heat. And summer was just beginning.

The temperature was rising, but the rain was not falling. The combination of record or near-record temperatures and low rainfall was drying out soils. Weekly drought maps published by the University of Nebraska showed drought-stricken areas spreading across more and more of the country until, by early July, these areas were engulfing virtually the entire Corn Belt. Soil moisture readings in the Corn Belt were among the lowest ever recorded.

While temperature, rainfall, and drought serve as indirect indicators of crop growing conditions, each week the U.S. Department of Agriculture releases a report on the actual state of the corn crop. This year the early reports were promising. On June 4th, 72 percent of the U.S. corn crop was rated as good to excellent—a strong early rating. But on June 11th the share of the crop in this category dropped to 66 percent. And then with each subsequent week it dropped further, until by July 9th only 40 percent of the U.S. corn crop was rated good to excellent. The other 60 percent was in very poor to fair condition. And the crop was still deteriorating.

Even during the few months when we were working on this book we were beginning to see how the more-extreme weather events that come with climate change can affect food security. Between the beginning of June and mid-July, corn prices increased by one third. Although the world was hoping for a good U.S. harvest to replenish dangerously low grain stocks, this will not likely happen.

World carryover stocks of grain will fall further at the end of this crop year, making the food situation even more precarious. Food prices, already elevated, will be climbing higher, quite possibly to record highs.

Not only is the current food situation deteriorating, so is the global food system itself. We saw early signs of the unraveling in 2008 following an abrupt doubling of world grain prices. As world food prices climbed, exporting countries began restricting exports to keep their domestic prices down. In response, governments of importing countries panicked. Some of them turned to buying or leasing land in other countries on which to produce food for themselves.

Welcome to the new geopolitics of food scarcity. As food supplies tighten, we are moving into a new food era, one in which it is every country for itself.

The world is in serious trouble on the food front. But there is little evidence that political leaders have yet grasped the magnitude of what is happening. The progress in reducing hunger in recent decades has been reversed. Feeding the world's hungry now depends on new population, energy, and water policies. Unless we move quickly to adopt new policies, the goal of eradicating hunger will remain just that.

The purpose of this book is to help people everywhere recognize that time is running out. The world may be much closer to an unmanageable food shortage—replete with soaring food prices, spreading food unrest, and ultimately political instability—than most people realize. This book is an effort by our Earth Policy research team to raise public understanding of the challenge that we are facing and to inspire action.

Lester R. Brown
July 2012

Earth Policy Institute
1350 Connecticut Ave. NW, Suite 403
Washington, DC 20036

Phone: (202) 496-9290
Fax: (202) 496-9325
epi@earth-policy.org
www.earth-policy.org

Full Planet, Empty Plates

1

Food:
The Weak Link

The world is in transition from an era of food abundance to one of scarcity. Over the last decade, world grain reserves have fallen by one third. World food prices have more than doubled, triggering a worldwide land rush and ushering in a new geopolitics of food. Food is the new oil. Land is the new gold.

The abrupt rise in world grain prices between 2007 and 2008 left more people hungry than at any time in history. It also spawned numerous food protests and riots. In Thailand, rice was so valuable that farmers took to guarding their ripened fields at night. In Egypt, fights in the long lines for state-subsidized bread led to six deaths. In poverty-stricken Haiti, days of rioting left five people dead and forced the Prime Minister to resign. In Mexico, the government was alarmed when huge crowds of tortilla protestors took to the streets.

After the doubling of world grain prices between 2007 and mid-2008, prices dropped somewhat during the recession, but this was short-lived. Three years later, high food prices helped fuel the Arab Spring.

We are entering a new era of rising food prices and spreading hunger. On the demand side of the food equation, population growth, rising affluence, and the conversion of food into fuel for cars are combining to raise consumption by record amounts. On the supply side, extreme soil erosion, growing water shortages, and the earth's rising temperature are making it more difficult to expand production. Unless we can reverse such trends, food prices will continue to rise and hunger will continue to spread, eventually bringing down our social system. Can we reverse these trends in time? Or is food the weak link in our early twenty-first-century civilization, much as it was in so many of the earlier civilizations whose archeological sites we now study?

This tightening of world food supplies contrasts sharply with the last half of the twentieth century, when the dominant issues in agriculture were overproduction, huge grain surpluses, and access to markets by grain exporters. During that time, the world in effect had two reserves: large carryover stocks of grain (the amount in the bin when the new harvest begins) and a large area of cropland idled under U.S. farm programs to avoid overproduction. When the world harvest was good, the United States would idle more land. When the harvest was subpar, it would return land to production. The excess production capacity was used to maintain stability in world grain markets. The large stocks of grain cushioned world crop shortfalls. When India's monsoon failed in 1965, for example, the United States shipped a fifth of its wheat harvest to India to avert a potentially massive famine. And because of abundant stocks, this had little effect on the world grain price.

When this period of food abundance began, the world had 2.5 billion people. Today it has 7 billion. From 1950 to 2000 there were occasional grain price spikes as a result of weather-induced events, such as a severe drought in Russia or an intense heat wave in the U.S. Midwest. But their effects on price were short-lived. Within a year or so things were back to normal. The combination of abundant stocks and idled cropland made this period one of the most food-secure in world history. But it was not to last. By 1986, steadily rising world demand for grain and unacceptably high budgetary costs led to a phasing out of the U.S. cropland set-aside program.

Today the United States has some land idled in its Conservation Reserve Program, but it targets land that is highly susceptible to erosion. The days of productive land ready to be quickly brought into production when needed are over.

Ever since agriculture began, carryover stocks of grain have been the most basic indicator of food security. The goal of farmers everywhere is to produce enough grain not just to make it to the next harvest but to do so with a comfortable margin. From 1986, when we lost the idled cropland buffer, through 2001, the annual world carryover stocks of grain averaged a comfortable 107 days of consumption.

This safety cushion was not to last either. After 2001, the carryover stocks of grain dropped sharply as world consumption exceeded production. From 2002 through 2011, they averaged only 74 days of consumption, a drop of one third. An unprecedented period of world food security has come to an end.

When world grain supplies tightened in 2007, there was no idled U.S. cropland to quickly return to production and there were no excess grain stocks to draw upon. Within two decades, the world had lost both of its safety cushions.

The world is now living from one year to the next, hoping always to produce enough to cover the growth in demand. Farmers everywhere are making an all-out effort to keep pace with the accelerated growth in demand, but they are having difficulty doing so.

Today the temptation for exporting countries to

restrict exports in order to dampen domestic food price rises is greater than ever. With another big jump in grain prices, we could see a breakdown in the world food supply system. If countries give in to the temptation to restrict exports, some lower-income importing countries might not be able to import any grain at all. When could this happen? We are not talking about the distant future. It could be anytime.

Food shortages undermined earlier civilizations. The Sumerians and Mayans are just two of the many early civilizations that declined apparently because they moved onto an agricultural path that was environmentally unsustainable. For the Sumerians, rising salt levels in the soil as a result of a defect in their otherwise well-engineered irrigation system eventually brought down their food system and thus their civilization. For the Mayans, soil erosion was one of the keys to their downfall, as it was for so many other early civilizations. We, too, are on such a path. While the Sumerians suffered from rising salt levels in the soil, our modern-day agriculture is suffering from rising carbon dioxide levels in the atmosphere. And like the Mayans, we too are mismanaging our land and generating record losses of soil from erosion.

While the decline of early civilizations can be traced to one or possibly two environmental trends such as deforestation and soil erosion that undermined their food supply, we are now dealing with several. In addition to some of the most severe soil erosion in human history, we are also facing newer trends such as the depletion of aquifers, the plateauing of grain yields in the more agriculturally advanced countries, and rising temperature.

Against this backdrop, it is not surprising that the U.N. Food Price Index was at 201 in June 2012, twice the base level of 100 in 2002–04. (See Figure 1–1.) For most Americans, who spend on average 9 percent of their income on

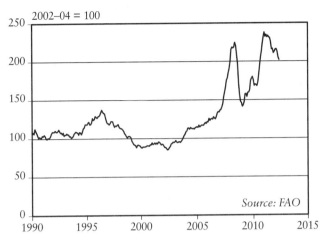

Figure 1–1. *World Monthly Food Price Index,*
January 1990–June 2012

food, this is not a big deal. But for consumers who spend 50–70 percent of their income on food, a doubling of food prices is a serious matter. There is little latitude for them to offset the price rise simply by spending more.

Closely associated with the decline in stocks of grain and the rise in food prices is the spread of hunger. During the closing decades of the last century, the number of hungry people in the world was falling, dropping to a low of 792 million in 1997. After that it began to rise, climbing toward 1 billion. Unfortunately, if we continue with business as usual, the ranks of the hungry will continue to expand.

Those trapped between low incomes and the doubling of world food prices are forced to eat less. Most of the nearly 1 billion people who are chronically hungry and malnourished live in the Indian subcontinent or sub-Saharan Africa. There are pockets of hunger elsewhere, but these are the two remaining regions where hunger is pervasive. India, which now has a thriving economy, should be experiencing a steady decline in the number who are hungry and malnourished. But it is not, presumably because rising incomes among the poor cannot keep up with rising food prices.

In a hungry world, it is children who suffer the most. Rising world food prices are leaving millions of children dangerously hungry. Some are too weak to walk to school. Many are so nutritionally deprived that they are physically and mentally stunted. Neither we nor they will ever know what their full human potential could be. The costs of this will be visible for decades to come.

As a result of chronic hunger, 48 percent of all children in India are stunted physically and mentally. They are undersized, underweight, and likely to have IQs that are on average 10–15 points lower than those of well-nourished children.

In early 2012, Adam Nossiter wrote in the *New York Times* about the effect of high food prices in the Democratic Republic of the Congo, a country where hunger is common. Interviewing individual families in Kinshasa, he noted that three years ago everyone ate at least one meal a day. But today even families with both parents working often cannot afford to eat every day. It is now a given in many households that some days will be foodless, days when they will not eat at all. Selecting the days when they will not eat is a weekly routine.

The international charity Save the Children commissioned detailed surveys in five countries—India, Pakistan, Nigeria, Peru, and Bangladesh—to see how people were dealing with rising food prices. Among other things, they learned that 24 percent of families in India now have foodless days. For Nigeria, the comparable figure is 27 percent. For Peru it is 14 percent. Family size plays an important role in hunger. Almost one third of large families in all countries surveyed have foodless days.

Historically there have been two sources of grain demand growth. The oldest of these is population growth. Each year the world adds nearly 80 million people. Tonight there will be 219,000 people at the dinner table who were not there last night, many of them with empty plates. Tomorrow night there will be another 219,000 people. Relentless population growth is putting excessive pressure on local land and water resources in many countries, making it difficult if not impossible for farmers to keep pace.

The second source of growing demand for grain is consumers moving up the food chain. As incomes rose in industrial countries after World War II, people began to consume more grain-intensive livestock and poultry products: meat, milk, and eggs. Today, with incomes rising fast in emerging economies, there are at least 3 billion people moving up the food chain in the same way. The largest single concentration of these new meat eaters is in China, which now consumes twice as much meat as the United States does.

Now there is a third source of demand for grain: the automobile. Distillers use grain to produce fuel ethanol for cars, an activity that is concentrated in the United States and that has developed largely since 2005. In 2011, the United States harvested nearly 400 million tons of grain. Of this, 127 million tons (32 percent) went to ethanol distilleries.

With this massive industrial capacity to convert grain into automotive fuel, the price of grain is now more closely linked to the price of oil than ever before. As the price of oil rises, it becomes more profitable to convert grain into ethanol. This sets the stage for competition for the grain harvest between the affluent owners of the world's 1 billion automobiles and the world's poorest people.

Population growth, the rising consumption of livestock and poultry products, and the use of grain to fuel cars together raised the world growth in grain consumption from an average of 21 million tons per year from 1990 to 2005 to 45 million tons per year from 2005 to 2011. Almost overnight, the annual growth in grain consumption doubled.

At a time when the world's farmers are facing this record growth in food demand, they continue to wrestle with the traditional threats to production such as soil erosion. But now they are also looking at three new challenges on the production front. One, aquifers are being depleted and irrigation wells are starting to go dry in 18 countries that together contain half the world's people. Two, in some of the more agriculturally advanced countries, rice and wheat yield per acre, which have been rising steadily for several decades, are beginning to plateau. And three, the earth's temperature is rising, threatening to disrupt world agriculture in scary ways.

The countries where water tables are falling and aquifers are being depleted include the big three grain producers—China, India, and the United States. World Bank data for India indicate that 175 million people are being fed with grain produced by overpumping. My own estimate for China is that 130 million people are being fed by overpumping. In the United States, the irrigated area is shrinking in leading agricultural states such as California and Texas as aquifers are depleted and irrigation water is diverted to cities.

Second, after several decades of rising grain yields, some of the more agriculturally advanced countries are hitting a glass ceiling, a limit that was not widely anticipated. Rice yields in Japan, which over a century ago became the first country to launch a sustained rise in land productivity, have not increased for 17 years. In both Japan and South Korea, yields have plateaued at just under 5 tons per hectare. (One hectare = 2.47 acres.) China's rice yields, rising rapidly in recent decades, are now closely approaching those of Japan. If China cannot raise its rice yields above those in Japan, and it does not seem likely that it can, then a plateauing there too is imminent.

A similar situation exists with wheat yields. In France, Germany, and the United Kingdom—the three leading wheat producers in Europe—there has been no rise for more than a decade. Other advanced countries will soon be hitting their glass ceiling for grain yields.

The third new challenge confronting farmers is global warming. The massive burning of fossil fuels is increasing the level of carbon dioxide in the atmosphere, raising the earth's temperature and disrupting climate. It is now in a state of flux. Historically when there was an extreme weather event—an intense heat wave or a drought—we knew it was temporary and that things would likely be back to normal by the next harvest. Now there is no "norm" to return to, leaving farmers facing a future fraught with risk.

High temperatures can lower crop yields. The widely used rule of thumb is that for each 1-degree-Celsius rise in temperature above the optimum during the growing season farmers can expect a 10-percent decline in grain yields. A historical study of the effect of temperature on corn and soybean yields in the United States found that a 1-degree-Celsius rise in temperature reduced grain yields 17 percent. Yet if the world continues with business as usual, failing to address the climate issue, the earth's temperature during this century could easily rise by 6 degrees Celsius (11 degrees Fahrenheit).

In recent years, world carryover stocks of grain have been only slightly above the 70 days that was considered a desirable minimum during the late twentieth century. Now stock levels must take into account the effect on harvests of higher temperatures, more extensive drought, and more intense heat waves. Although there is no easy way to precisely quantify the harvest effects of any of these climate-related threats, it is clear that any of them can shrink harvests, potentially creating chaos in the world grain market. To mitigate this risk, a stock reserve equal to 110 days of consumption would produce a much safer level of food security.

Although we talk about food price spikes, what we are more likely starting to see is a ratcheting upward of food prices. This process is likely to continue until we succeed in reversing some of the trends that are driving it. All of the threatening trends are of human origin, but whether we can reverse them remains to be seen.

As food supplies tighten, the geopolitics of food is fast overshadowing the geopolitics of oil. The first signs of trouble came in 2007, when world grain production fell behind demand. Grain and soybean prices started to climb, doubling by mid-2008. In response, many exporting countries tried to curb rising domestic food prices by restricting exports. Among them were Russia and Argentina, two leading wheat exporters. Viet Nam, the world's number two rice exporter, banned exports entirely in the early months of 2008. Several other smaller grain suppliers also restricted exports.

With key suppliers restricting or banning exports, importing countries panicked. No longer able to rely on the market for grain, several countries tried to negotiate long-term grain supply agreements with exporting countries. The Philippines, a chronically rice-deficit country, attempted to negotiate a three-year agreement with Viet Nam for 1.5 million tons of rice per year. A delegation of Yemenis traveled to Australia with a similar goal in mind for wheat, but they had no luck. In a seller's market, exporters were reluctant to make long-term commitments.

Fearing they might not be able to buy needed grain

from the market, some of the more affluent countries, led by Saudi Arabia, China, and South Korea, then took the unusual step of buying or leasing land long term in other countries on which to grow food for themselves. These land acquisitions have since grown rapidly in number. Most of them are in Africa. Among the principal destinations for land hunters are Ethiopia, Sudan, and South Sudan, each of them countries where millions of people are being sustained with food donations from the U.N. World Food Programme.

As of mid-2012, hundreds of land acquisition deals had been negotiated or were under negotiation, some of them exceeding a million acres. A 2011 World Bank analysis of these "land grabs" reported that at least 140 million acres were involved—an area that exceeds the cropland devoted to corn and wheat combined in the United States. This onslaught of land acquisitions has become a land rush as governments, agribusiness firms, and private investors seek control of land wherever they can find it. Such acquisitions also typically involve water rights, meaning that land grabs potentially affect downstream countries as well. Any water extracted from the upper Nile River basin to irrigate newly planted crops in Ethiopia, Sudan, or South Sudan, for instance, will now not reach Egypt, upending the delicate water politics of the Nile by adding new countries that Egypt must compete with for water.

The potential for conflict is high. Many of the land deals have been made in secret, and much of the time the land involved was already being farmed by villagers when it was sold or leased. Often those already farming the land were neither consulted nor even informed of the new arrangements. And because there typically are no formal land titles in many developing-country villages, the farmers who lost their land have had little support for bringing their cases to court.

The bottom line is that it is becoming much more difficult for the world's farmers to keep up with the world's rapidly growing demand for grain. World grain stocks were drawn down a decade ago and we have not been able to rebuild them. If we cannot do so, we can expect that with the next poor harvest, food prices will soar, hunger will intensify, and food unrest will spread. We are entering a time of chronic food scarcity, one that is leading to intense competition for control of land and water resources—in short, a new geopolitics of food.

Data, endnotes, and additional resources can be found at Earth Policy Institute, www.earth-policy.org.

2

The Ecology of Population Growth

Throughout most of human existence, population growth has been so slow as to be imperceptible within a single generation. Reaching a global population of 1 billion in 1804 required the entire time since modern humans appeared on the scene. To add the second billion, it took until 1927, just over a century. Thirty-three years later, in 1960, world population reached 3 billion. Then the pace sped up, as we added another billion every 13 years or so until we hit 7 billion in late 2011.

One of the consequences of this explosive growth in human numbers is that human demands have outrun the carrying capacity of the economy's natural support systems—its forests, fisheries, grasslands, aquifers, and soils. Once demand exceeds the sustainable yield of these natural systems, additional demand can only be satisfied by consuming the resource base itself. We call this over-cutting, overfishing, overgrazing, overpumping, and over-plowing. It is these overages that are undermining our global civilization.

The exponential growth that has led to this explosive increase in our numbers is not always an easy concept to grasp. As a result, not many of us—including political leaders—realize that a 3 percent annual rate of growth will actually lead to a 20-fold growth in a century.

The French use a riddle to teach exponential growth to schoolchildren. A lily pond, so the riddle goes, contains a single leaf. Each day the number of leaves doubles— two leaves the second day, four the third, eight the fourth, and so on. Question: "If the pond is full on the thirtieth day, at what point is it half full?" Answer: "On the twenty-ninth day." Our global lily pond may already be in the thirtieth day.

The most recent U.N. demographic projections show world population growing to 9.3 billion by 2050, an addition of 2.3 billion people. Most people think these demographic projections, like most of those made over the last half-century, will in fact materialize. But this is unlikely, given the difficulties in expanding the food supply, such as those posed by spreading water shortages and global warming. We are fast outgrowing the earth's capacity to sustain our increasing numbers.

World population growth has slowed from the peak of 2.1 percent in 1967 to 1.1 percent in 2011. What is not clear is whether population growth will slow further because we accelerate the shift to smaller families or because we fail to do so and eventually death rates begin to rise. We know what needs to be done. Millions of women in the world want to plan their families but lack access to reproductive health and family planning services. Filling this gap would not only take us a long way toward stabilizing world population, it would also improve the health and well-being of women and their families.

Population projections are based on numerous demographic assumptions, including, among others, fertility levels, age distribution, and life expectancy. They sometimes create the illusion that the world can support these huge increases. But demographers rarely ask such questions as, Will there be enough water to grow food for 2.3 billion more people? Will population growth continue without interruption in the face of crop-shrinking heat waves?

As human numbers multiply, we need more and more irrigation water. As a result, half of the world's people now live in countries that are depleting their aquifers by overpumping. Overpumping is by definition a short-term phenomenon.

The situation is similar with fishing, as world population growth has increased demand for seafood. A fishing fleet can continue expanding the fish catch until it exceeds the reproductive capacity of a fishery. When this happens, the fishery begins to shrink and eventually collapses. A startling 80 percent of oceanic fisheries are being fished at or beyond their sustainable yield.

When oceanic fisheries collapse, we turn to fish farming. Doing this, however, takes land and water, since these fish must be fed, most often with some combination of corn and soybean meal. Thus, collapsing fisheries put additional pressure on the earth's land and water resources.

As human populations grow, so typically do livestock populations, particularly in those parts of the world where herding cattle, sheep, and goats is a way of life. This is most evident in Africa, where the explosion in human numbers from 294 million in 1961 to just over 1 billion in 2010 was accompanied by growth in the livestock population from 352 million to 894 million.

With livestock numbers growing beyond the sustainable yield of grasslands, these ecosystems are deteriorating. The loss of vegetation leaves the land vulnerable to soil erosion. At some point, the grassland turns to desert, depriving local people of their livelihood and food supply, as is now happening in parts of Africa, the Middle East, central Asia, and northern China.

Growing populations also increase the demand for firewood, lumber, and paper. The result is that demand for wood is exceeding the regenerative capacity of forests. The world's forests, which have been shrinking for several decades, are currently losing a net 5.6 million hectares per year. In the absence of a more responsible population policy, forested area will continue to shrink. Some countries—Mauritania is one example—have lost nearly all their forest and are now essentially treeless. Without trees to protect the soil and to reduce runoff, the entire ecosystem suffers, making it more difficult to produce enough food.

Continuous population growth eventually leads to overplowing—the breaking of ground that is highly erodible and should not be plowed at all. We are seeing this in Africa, the Middle East, and much of Asia. Plowing marginal land leads to soil erosion and eventually to cropland abandonment. Land that would otherwise sustain grass and trees is lost as it is converted into cropland and then turns into wasteland.

In summary, we have ignored the earth's environmental stop signs. Faced with falling water tables, not a single country has mobilized to reduce water use so that it would not exceed the sustainable yield of an aquifer. Unless we can stop willfully ignoring the threats and wake up to the risks we are taking, we will join the earlier civilizations that failed to reverse the environmental trends that undermined their food economies.

The good news is that 44 countries, including nearly all those in both Western and Eastern Europe, have reached population stability as a result of gradual fertility decline over the last several generations. Their populations total 970 million people, roughly one seventh of humanity.

Two other geographic regions are now moving rapidly toward population stability. East Asia, including Japan, North and South Korea, China, and Taiwan, a region of over 1.5 billion people, is very close to stabilizing its population. Japan's population is already declining. The populations of the two Koreas and Taiwan are still growing, but slowly. China's population of 1.35 billion is projected to peak in 2026 at 1.4 billion and then start shrinking. By 2045 its population will likely be smaller than it is today.

In Latin America, a combination of poverty reduction and broad access to family planning services is slowing population growth. Its population of just over 600 million in 2012 is projected to reach 751 million by 2050. Brazil, by far the largest country in the region, is projected to expand from 198 million in 2012 to 223 million in 2050, a growth of only 12 percent over nearly four decades.

The bad news in our demographic future is that virtually all of the population growth will be in developing countries, the areas least able to support them. The two regions where most future population growth will occur are the Indian subcontinent and sub-Saharan Africa. The Indian subcontinent, principally India, Pakistan, and Bangladesh, which now has nearly 1.6 billion people, is projected to reach almost 2.2 billion by 2050. Africa south of the Sahara, with 899 million people today, also is projected to hit 2.2 billion by 2050. The big challenge for the world today is to help countries in these two regions accelerate the shift to smaller families, both by eradicating poverty and by ensuring that all women have access to reproductive health care and family planning services, thus avoiding stressful growth in population.

The contrast between countries that have essentially stabilized their populations and those where large families are still the rule could not be greater. On one end of the spectrum are Germany with 82 million people, Russia with 143 million, and Japan with 126 million. Populations in all three are projected to shrink by roughly one tenth by 2050. With elderly populations and low birth rates, deaths now exceed births in each of these countries.

Meanwhile, Nigeria, Ethiopia, and Pakistan are anticipating massive growth. Nigeria, geographically not much larger than Texas, now has 167 million people and is projected to have 390 million by 2050. In Ethiopia, one of the world's hungriest countries, the current population of 87 million is expected to reach 145 million by 2050. And Pakistan, with 180 million people living on the equivalent of 8 percent of the U.S. land area, is projected to reach 275 million by 2050—nearly as many people as in the United States today.

The "demographic transition" helps us understand what happens to population growth in individual countries as they develop. In 1945, Princeton demographer Frank Notestein outlined a three-stage demographic model to illustrate the dynamics of population growth as societies modernized. He pointed out that in pre-modern societies, where both births and deaths are high, there is little or no population growth. In stage two, as living standards rise and health care improves, death rates begin to decline. With birth rates remaining high while death rates are declining, population growth accelerates, typically reaching close to 3 percent a year. As living standards continue to improve, and particularly as women are educated, the birth rate also begins to decline. Eventually the birth rate drops to the level of the death rate. This is stage three of the demographic transition, where births and deaths are in balance and population is again stable.

Most countries have made it at least as far as stage two, while many industrialized countries have long since reached stage three. Sadly, many countries have not been able to lower their birth rates to make it into stage three. Stage two becomes a demographic trap for them. Their populations are growing continuously at 3 percent a year—a rate that, as mentioned earlier, leads to a 20-fold increase in a century. For example, if the 2012 population of Tanzania, one of Africa's larger countries, of nearly 48 million continued to grow at 3 percent a year, the country would have 916 million people within a hundred years. Iraq's population of 34 million, also expanding at 3 percent a year, would reach 648 million a century hence.

Governments in countries that have experienced such rapid population growth for two generations are showing signs of demographic fatigue. Worn down by the struggle to build schools and provide jobs for an ever-expanding population, they are facing political stresses on every hand.

Countries that fail to shift to smaller families risk being overwhelmed by land and water shortages, disease, civil conflict, and other adverse effects of prolonged rapid population growth. We call them failing states—countries where governments can no longer provide personal security, food security, or basic social services such as education and health care. Governments lose their legitimacy and often their authority to govern. Countries in this situation include Yemen, Ethiopia, Somalia, the Democratic Republic of the Congo, and Afghanistan. Among the more populous failing states are Pakistan and Nigeria.

Based on a Fund for Peace list published each year in *Foreign Policy* magazine, the top 20 failing states, almost without exception, have high levels of fertility. In Afghanistan and Somalia, for example, women have on average six children. These countries demonstrate how population growth and state disintegration can reinforce each other.

The countries that have made it into stage three, with lower fertility and fewer children, benefit from higher rates of savings. They are reaping what economic demographers call the "demographic bonus." When a country shifts quickly to smaller families, the number of young dependents—those who need nurturing and educating— declines sharply relative to the number of working adults.

As household savings climb, investment rises and economic growth accelerates.

Virtually all countries that have quickly shifted to smaller families have benefited from this bonus. After World War II, Japan made a concerted effort to slow its population growth, cutting its growth rate in half between 1948 and 1955. It became the first country to gain the bonus benefit. The spectacular economic growth over the next three decades, unprecedented in any country, raised Japan's income per person to one of the highest in the world, making it a modern industrial economy that was second in size only to the United States.

South Korea, Taiwan, Hong Kong, and Singapore followed shortly thereafter. These four so-called tiger economies, which enjoyed such spectacular economic growth during the late twentieth century, each benefited from a rapid fall in birth rates and the demographic bonus that followed.

On a much larger scale, China's declining birth rate, mainly a result of its one-child family program, created an unusually large demographic bonus, helping people save a good share of their incomes and thus spurring investment. The phenomenal investment rate, coupled with the record influx of private foreign investment and accompanying technology, is fast propelling China into the ranks of modern industrial powers. Other countries with age structures now conducive to high savings and rapid economic growth include Sri Lanka, Mexico, Iran, Tunisia, and Viet Nam.

We all have a stake in ensuring that countries everywhere move into stage three of the demographic transition. Those that are caught in the demographic trap are likely to be politically unstable—often overcome by internal conflict. These failing states are more likely to be breeding grounds for terrorists than to be participants in building a stable world order.

If world population growth does not slow dramatically, the number of people trapped in hydrological poverty and hunger will almost certainly grow, threatening food security, economic progress, and political stability. The only humane option is to move quickly to replacement-level fertility of two children per couple and to stabilize world population as soon as possible.

Data, endnotes, and additional resources can be found at Earth Policy Institute, www.earth-policy.org.

3

Moving Up the Food Chain

For most of the time that human beings have walked the earth, we lived as hunter-gatherers. The share of the human diet that came from hunting versus gathering varied with geographic location, hunting skills, and the season of the year. During the northern hemisphere winter, for instance, when there was little food to gather, people there depended heavily on hunting for survival. Our long history as hunter-gatherers left us with an appetite for animal protein that continues to shape diets today.

As recently as the closing half of the last century, a large part of the growth in demand for animal protein was still satisfied by the rising output of two natural systems: oceanic fisheries and rangelands. Between 1950 and 1990, the oceanic fish catch climbed from 17 million to 84 million tons, a nearly fivefold gain. During this period, the seafood catch per person more than doubled, climbing from 15 to 35 pounds.

This was the golden age of oceanic fisheries. The catch grew rapidly as fishing technologies evolved and as refrigerated processing ships began to accompany fishing fleets, enabling them to operate in distant waters. Unfortunately, the human appetite for seafood has outgrown the sustainable yield of oceanic fisheries. Today four fifths of fisheries are being fished at or beyond their sustainable capacity. As a result, many are in decline and some have collapsed.

Rangelands are also essentially natural systems. Located mostly in semiarid regions too dry to sustain agriculture, they are vast—covering roughly twice the area planted to crops. In some countries, such as Brazil and Argentina, beef cattle are almost entirely grass-fed. In others, such as the United States and those in Europe, beef is produced with a combination of grass and grain.

In every society where incomes have risen, the appetite for meat, milk, eggs, and seafood has generated an enormous growth in animal protein consumption. Today some 3 billion people are moving up the food chain. For people living at subsistence level, 60 percent or more of their calories typically come from a single starchy food staple such as rice, wheat, or corn. As incomes rise, diets are diversified with the addition of more animal protein.

World consumption of meat climbed from just under 50 million tons in 1950 to 280 million tons in 2010, more than a fivefold increase. Meanwhile, consumption per person went from 38 pounds to 88 pounds a year. The growth in consumption during this 60-year span was concentrated in the industrial and newly industrializing countries.

The type of animal protein that people choose to eat depends heavily on geography. Countries that are land-rich with vast grasslands—including the United States, Brazil, Argentina, and Russia—depend heavily on beef or—as in Australia and Kazakhstan—mutton. Countries that are more densely populated and lack extensive grazing lands have historically relied much more on pork. Among these are Germany, Poland, and China. Island countries and those with long shorelines, such as Japan and Norway, have turned to the oceans for their animal protein.

Over time, global patterns of meat consumption have changed. In 1950, beef and pork totally dominated, leaving poultry a distant third. From 1950 until 1980, beef and pork propduction increased more or less apace. Beef production was pressing against the limits of grasslands, however, and more cattle were put in feedlots. Because cattle are not efficient in converting grain into meat, world beef production, which climbed from 19 million tons in 1950 to 53 million in 1990, has not expanded much since then. In contrast, chickens are highly efficient in converting grain into meat. As a result, world poultry production, which grew slowly at first, accelerated, overtaking beef in 1997. (See Figure 3–1.)

The world's top two meat consumers are China and the United States. The United States was the leader until 1992, when it was overtaken by China. (See Figure 3–2.) As of 2012, twice as much meat is eaten in China as in the United States—71 million tons versus 35 million.

The huge growth in meat consumption in China, mostly of pork, came after the economic reforms in 1978, when

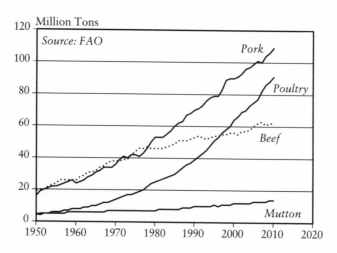

Figure 3–1. *World Meat Production by Type, 1950–2010*

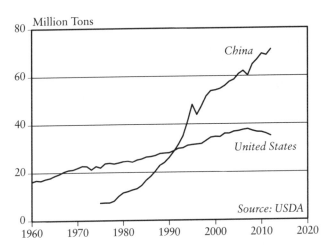

Figure 3–2. *Meat Consumption in China
and the United States, 1960–2012*

large production teams were replaced by family farms. Today pork is the world's leading meat, and half of it is eaten in China. The heavy reliance on pork in China is not new. In an effort to minimize waste, village families in China have a long-standing tradition of keeping a pig that is fed kitchen and table wastes. When the pig matures, it is butchered and eaten and replaced with another small, recently weaned, pig. Even though large-scale commercial hog production now dominates output in urbanizing China, pork's prominent place in the Chinese diet has deep cultural roots.

With China's 1.35 billion people clamoring for more pork, production there climbed from 9 million tons in 1978, the year of the economic reforms, to 52 million tons in 2012. U.S. pork production rose from 6 million to 8 million tons during the same period.

These shifts in world meat consumption have been driven primarily by widely differing production costs, with consumers moving toward the lower-cost offerings. In 1950, poultry was expensive and production was limited, roughly the same as mutton. But from mid-century onward, advances in the efficiency of poultry production dropped the price to where more and more people could afford it. In the United States—where a half-century ago it was something special, usually served at Sunday dinner—the low price of chicken now makes it the meat of choice for everyday consumption.

Perhaps the greatest restructuring is occurring with seafood consumption. Historically, as the demand for seafood increased and fishing technologies advanced, coastal and island countries in particular began to rely more heavily on the oceans. As population pressure built up in Japan, more and more land was needed to produce its food staple, rice. By the early twentieth century, Japan was using virtually all its arable land to produce rice, leaving none to produce feed for livestock and poultry. So Japan turned to seafood to satisfy the growing demand for animal protein.

Japan now consumes 8 million tons of seafood a year as part of its "fish and rice" diet. But with oceanic fisheries being pushed to their limits, there are few opportunities for other countries to turn seaward for protein in the same way. For example, if China's per capita consumption of oceanic seafood were to reach the Japanese level, it would consume nearly the entire world catch.

So although China is a leading claimant on oceanic fisheries, with an annual catch of 15 million tons, it has turned primarily to fish farming to meet its fast-growing seafood needs. As of 2010, its aquacultural output—mainly carp and shellfish—totaled 37 million tons, more than the rest of the world combined. With incomes now rising in densely populated Asia, other countries are following China's lead. Among them are India, Thailand, and Viet Nam.

Over the last 20 years, aquaculture has thus emerged as a major source of animal protein. Driven by the high efficiency with which omnivorous species such as carp, tilapia, and catfish convert grain into animal protein, world aquacultural output expanded more than fourfold between 1990 and 2010. (See Figure 3–3.) Early estimates indicate it eclipsed beef production worldwide in 2011.

Not all aquacultural operations are environmentally beneficial. Some are both environmentally disruptive and inefficient in feed use, such as the farming of shrimp and salmon. These operations account for only a small share of the global farmed fish total, but they are growing fast. Shrimp farming often involves the destruction of coastal mangrove forests to create habitat for the shrimp. Salmon are inefficient in that they are fed other fish, usually as fish-meal, which comes either from fish processing plant wastes or from low-value fish caught specifically for this purpose.

As people consume more meat, milk, eggs, and farmed

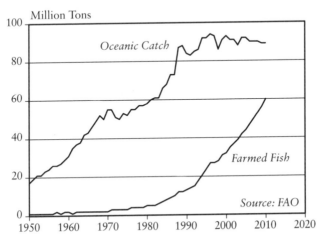

Figure 3–3. *World Oceanic Fish Catch and Farmed Fish Production, 1950–2010*

fish, indirect grain consumption rises. Comparing grain use per person in India and the United States provides some idea of how much grain it takes to move up the food chain. In low-income India—where annual grain consumption totals 380 pounds per person, or roughly 1 pound a day—nearly all grain must be eaten directly to satisfy basic food energy needs. Only 4 percent is converted into animal protein. Not surprisingly, the consumption of most livestock products in India is rather low. Milk, egg, and poultry consumption, however, are beginning to rise, particularly among India's expanding middle class.

The average American, in contrast, consumes roughly 1,400 pounds of grain per year, four fifths of it indirectly in the form of meat, milk, and eggs. Thus the total grain consumption per person in the United States is nearly four times that in India.

Pork and poultry meat are the world's leading sources of land-based animal protein, but eggs are not far behind, with 69 million tons produced in 2010. Egg production has grown steadily over the last half-century and appears likely to continue to do so. Eggs are a relatively inexpensive but valuable serving size source of protein. Worldwide, people on average eat three eggs per week.

As with pork, egg production in China has grown at an explosive pace, going from 6 million tons in 1990 to 24 million tons in 2010. As a result, China totally dominates world egg production. The United States is a distant second, with just over 5 million tons per year. India ranks third, with 3 million tons.

Yet consumers in some countries live high on the food chain but use relatively little grain to feed animals. For example, the Japanese use only moderate amounts of feedgrains because their protein intake is dominated by the oceanic fish catch. This is also the case with Argentina and Brazil, where nearly all the beef is grass-fed.

In recent decades, Brazil, the world's third ranking meat consumer, has experienced a marked restructuring of its meat consumption pattern. In 1960 beef was totally dominant, with pork a distant second and poultry almost nonexistent. By 2000, to the surprise of many, the fast-growing consumption of poultry in Brazil eclipsed that of beef. Pork consumption is still far behind.

With the world's grasslands being grazed at their limits or beyond, additional beef production now comes largely from putting more cattle in feedlots. A steer in a feedlot requires 7 pounds of grain for each pound of weight gain. For pork, each pound of additional live weight requires 3.5 pounds. For poultry, it is just over 2. For eggs the ratio is 2 to 1. For carp in China and India and catfish in the United States, it takes less than 2 pounds of feed for each pound of additional weight gain. Thus the worldwide change in patterns of meat consumption reflects the costs of meat production, which in turn reflects the widely varying levels of efficiency with which cattle, pigs, chickens, and farmed fish convert grain into protein.

Recent production trends give some sense of where the world is headed. Between 1990 and 2010, growth in beef production averaged less than 1 percent a year. Pork, meanwhile, expanded at over 2 percent annually, eggs at nearly 3 percent, and poultry at 4 percent. Aquacultural output, which sets the gold standard in grain conversion efficiency, expanded by nearly 8 percent a year, climbing from 13 million tons in 1990 to 60 million tons in 2010.

The share of the world grain harvest used for feeding livestock, poultry, and farmed fish has remained remarkably stable over the last few decades. One reason it has not risen much is the practice, now worldwide, of incorporating soybean meal into feed rations at a ratio of roughly 1 part soybean meal to 4 parts grain. This leads to a much more efficient conversion of grain into animal protein.

As the demand for animal protein has climbed over the last half-century, demand for soybeans has climbed even faster. (See Chapter 9.)

Worldwide, roughly 35 percent of the 2.3-billion-ton annual grain harvest is used for feed. In contrast, nearly all of the soybean harvest ends up as feed. Both pork and poultry output depend heavily on grain, whereas beef and milk production depend more on a combination of grass and grain.

The world's three largest meat producers—China, the United States, and Brazil—rely heavily on soybean meal as a protein supplement in feed rations. Indeed, the share of soybean meal in feed in each country now ranges between 15 and 18 percent.

The mounting pressure on land and water resources has led to some promising new animal protein production models, one of which is milk production in India. Since 1970, India's milk production has increased nearly sixfold, jumping from 21 million to 117 million tons. In 1997, India overtook the United States in dairy production, making it the world's leading milk producer.

The spark for this explosive growth came in 1965 when an enterprising young Indian, Dr. Verghese Kurien, organized the National Dairy Development Board, an umbrella organization of dairy cooperatives. The co-op's principal purpose was to market the milk from the two or three cows typically owned by each village family. It was these dairy cooperatives that provided the link between the growing appetite for dairy products and the millions of village families who had only a small marketable surplus.

Creating the market for milk spurred the sixfold growth in output. In a country where protein shortages stunt the growth of so many children, expanding the milk supply from less than half a cup per person a day 25 years ago to more than a cup today represents a major advance.

What is unique here is that India has built the world's largest dairy industry almost entirely on roughage, mostly crop residues—wheat straw, rice straw, and corn stalks— and grass collected from the roadside. Cows are often stall-fed with crop residues or grass gathered daily and brought to them.

A second relatively recent protein production model, which also relies on ruminants, is one developed in China, principally in four provinces of central eastern China— Hebei, Shangdong, Henan, and Anhui—where double cropping of winter wheat and corn is common. Once the winter wheat matures and ripens in early summer, it must be harvested quickly so that the seedbed can be prepared for corn planting. The straw that is removed from the land prior to preparing the seedbed is fed to cattle, as are the cornstalks left after the corn harvest in late fall. By supplementing this roughage with small amounts of nitrogen, typically in the form of urea, the microflora in the complex four-stomach digestive system of cattle can convert roughage efficiently into animal protein.

This practice enables these four crop-producing provinces to produce much of the country's beef as well. This central eastern region of China, dubbed the Beef Belt by Chinese officials, is producing large quantities of animal protein using only roughage. This use of crop residues to produce milk in India and beef in China means farmers are reaping a second harvest from the original crop.

Another highly efficient animal protein production model, one that has evolved in China over the centuries, is found in aquaculture. In a carp polyculture production system, four species of carp are grown together. One species feeds on phytoplankton. One feeds on zooplankton. A third feeds on aquatic grass. And the fourth is a bottom feeder. These four species thus form a small ecosystem, with each filling a particular niche. This multispecies system accounts for the major part of China's carp harvest of 16 million tons in 2011.

Although these three protein production models have evolved in India and China, both densely populated nations, they may find a place in other parts of the world as population pressures intensify and as people seek new ways to convert plant products into animal protein.

Looking to the future, there are some rather obvious shifts occurring in the pattern of world meat consumption. These are largely driven by an ongoing shift from the less efficient converters of grain into animal protein, such as feedlot beef, toward the more efficient converters, such as farmed fish and poultry. If recent trends continue, poultry production, which has already eclipsed beef, will likely overtake pork in 2020 or shortly thereafter, making poultry the world's leading meat. Within a few years, the production of farmed fish is likely to overtake both poultry and pork, becoming the world's leading source of animal protein by 2023.

In the United States, meat consumption, which had climbed steadily for over half a century, peaked in 2007, dropping 6 percent by 2012. This peak and decline were not widely anticipated. Among the contributing factors are high feed prices and, hence, meat prices; lingering uncertainty by consumers about the economic recovery; and a growing awareness among consumers of the negative health consequences of eating too much meat, including heart disease, cancer, and obesity. There is also growing opposition by animal rights and environmental groups to the inhumane production methods and pollution associated with factory farming. For one reason or another, Americans are reducing their consumption of meat. The United States seems to be the first among the more populous countries to experience such an abrupt decline—one that appears likely to become a longer-term trend.

People with the longest life expectancy are not those who live very low or very high on the food chain but those who occupy an intermediate position. Italians, who live lower on the food chain than Americans do, can expect to live for 81 years, compared with American life expectancy of 79. Italians benefit from what is commonly described as the Mediterranean diet, one that includes livestock and poultry but in moderate amounts.

Although the world has had many years of experience in feeding nearly 80 million more people each year, it has much less experience with also providing for 3 billion people with rising incomes who want to move up the food chain and consume more grain-intensive products. Whereas population growth generates demand for wheat and rice, humanities' two food staples, it is rising affluence that is driving growth in the demand for corn, the world's feedgrain. Historically, world corn and wheat production trends moved more or less together from 1950 until 2000. But then corn took off, climbing to 960 million tons in 2011 while wheat remained under 700 million tons.

It is the increase in consumption of livestock products plus the conversion of grain into fuel that have boosted the annual growth in world grain demand from the roughly 20 million tons a decade ago to over 40 million tons in recent years. As incomes continue to rise, the pressure on farmers to produce enough grain and soybeans to satisfy the growing appetite for livestock and poultry products will only intensify.

Data, endnotes, and additional resources can be found at Earth Policy Institute, www.earth-policy.org.

4

Food or Fuel?

At the time of the Arab oil export embargo in the 1970s, the importing countries were beginning to ask themselves if there were alternatives to oil. In a number of countries, particularly the United States, several in Europe, and Brazil, the idea of growing crops to produce fuel for cars was appealing. The modern biofuels industry was launched.

This was the beginning of what would become one of the great tragedies of history. Brazil was able to create a thriving fuel ethanol program based on sugarcane, a tropical plant. Unfortunately for the rest of the world, however, in the United States the feedstock was corn. Between 1980 and 2005, the amount of grain used to produce fuel ethanol in the United States gradually expanded from 1 million to 41 million tons.

Then came Hurricane Katrina, which disrupted Gulf-based oil refineries and gasoline supply lines in late August 2005. As gasoline prices in the United States quickly climbed to $3 a gallon, the conversion of a $2 bushel of corn, which can be distilled into 2.8 gallons of ethanol, became highly profitable.

The result was a rush to raise capital and build distilleries. From November 2005 through June 2006, ground was broken for a new ethanol plant in the United States every nine days. From July through September, the construction pace accelerated to one every five days. And in October 2006, it was one every three days.

Between 2005 and 2011, the grain used to produce fuel for cars climbed from 41 million to 127 million tons— nearly a third of the U.S. grain harvest. (See Figure 4–1.) The United States is trying to replace oil fields with corn fields to meet part of its automotive fuel needs.

The massive diversion of grain to fuel cars has helped drive up food prices, leaving low-income consumers everywhere to suffer some of the most severe food price inflation in history. As of mid-2012, world wheat, corn, and soybean prices were roughly double their historical levels.

The appetite for grain to fuel cars is seemingly insatiable. The grain required to fill a 25-gallon fuel tank of a

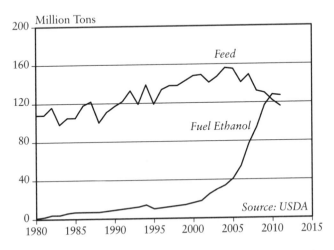

Figure 4–1. *Corn Use for Feed and Fuel Ethanol in the United States, 1980–2011*

sport utility vehicle with ethanol just once would feed one person for a whole year. The grain turned into ethanol in the United States in 2011 could have fed, at average world consumption levels, some 400 million people. But even if the entire U.S. grain harvest were turned into ethanol, it would only satisfy 18 percent of current gasoline demand.

With its enormous growth in distilling capacity, the United States quickly overtook Brazil to become the new world leader in biofuels. In 2011, the United States produced 14 billion gallons of ethanol and Brazil produced under 6 billion gallons; together they accounted for 87 percent of world output. The 14 billion gallons of U.S. grain-based ethanol met roughly 6 percent of U.S. gasoline demand. Other countries producing ethanol from food crops, though in relatively small amounts, include China, Canada, France, and Germany.

Most ethanol production growth has been concentrated in the last several years. In 1980, the world produced scarcely 1 billion gallons of fuel ethanol. By 2000, the figure was 4.5 billion gallons. It was still increasing, albeit slowly, expanding to 8.2 billion gallons in 2005. But between then and 2011, production jumped to 23 billion gallons.

A number of countries, including the United States, are also producing biodiesel from oil-bearing crops. World biodiesel production grew from a mere 3 million gallons in 1991 to just under 1 billion gallons in 2005. During the next six years it jumped to nearly 6 billion gallons, increasing sixfold. Still, worldwide production of biodiesel is less than one fourth that of ethanol.

The production of biodiesel is much more evenly distributed among countries than that of ethanol. The top five producers are the United States, Germany, Argentina, Brazil, and France, with production ranging from 840 million gallons per year in the United States to 420 million gallons in France.

A variety of crops can be used to produce biodiesel. In Europe, where sunflower seed oil, palm oil, and rapeseed oil are leading table oils, rapeseed is used most often for biodiesel. Similarly, in the United States the soybean is the leading table oil and biodiesel feedstock. Elsewhere, palm oil is widely used both for food and to produce biodiesel.

Although production from oil palms is limited to tropical and subtropical regions, the crop yields much more biodiesel per acre than do temperate-zone oilseeds such as soybeans and rapeseed. However, one disturbing consequence of rising biofuel production is that new oil palm plantations are coming at the expense of tropical forests. And any land that is devoted to producing biofuel crops is not available to produce food.

Not only are biofuels helping raise food prices, and thus increasing the number of hungry people, most make little sense from an energy efficiency perspective. Although ethanol can be produced from any plant, it is much more efficient and much less costly to use sugar- and starch-bearing crops. But even among these crops the efficiency varies widely. The ethanol yield per acre from sugarcane is nearly 600 gallons, a third higher than that from corn. This is partly because sugarcane is grown in tropical and subtropical regions and it grows year-round. Corn, in contrast, has a growing season of 120 days or so.

In terms of energy efficiency, grain-based ethanol is a clear loser. For sugarcane, the energy yield—that is, the energy embodied in the ethanol—can be up to eight times the energy invested in producing the biofuel. In contrast, the energy return on energy invested in producing corn-based ethanol is only roughly 1.5 to 1, a dismal return.

For biodiesel, oil palm is far and away the most energy-efficient crop, yielding roughly nine times as much energy as is invested in producing biodiesel from it. The energy return for biodiesel produced from soybeans and rapeseed is about 2.5 to 1. In terms of land productivity, an acre of oil palms can produce over 500 gallons of fuel per year—more than six times that produced from soybeans or rapeseed. Growing even the most productive fuel crops, however, still means either diverting land from other crops or clearing more land.

The capacity to convert enormous volumes of grain into fuel means that the price of grain is now more closely tied to the price of oil than ever before. If the price of fuel from grain drops below that from oil, then investment in converting grain into fuel will increase. Thus, if the price of oil were to reach, say, $200 a barrel, there would likely be an enormous additional investment in ethanol distilleries to convert grain into fuel. If the price of corn rises high enough, however, distilling grain to produce fuel may no longer be profitable.

One of the consequences of integrating the world food and fuel economies is that the owners of the world's 1 billion motor vehicles are pitted against the world's poorest people in competition for grain. The winner of this competition will depend heavily on income levels. Whereas the average motorist has an annual income over $30,000, the incomes of the 2 billion poorest people in the world are well under $2,000.

Rising food prices can quickly translate into social unrest. As grain prices were doubling from 2007 to mid-2008, food protests and riots broke out in many countries. Economic stresses in the form of rising food prices are translating into political stresses, putting governments in some countries under unmanageable pressures. The U.S. State Department reports food unrest in some 60 countries between 2007 and 2009. Among these were Afghanistan, Yemen, Ethiopia, Somalia, Sudan, the Democratic Republic of the Congo, and Haiti.

International food assistance programs are also hit

hard by rising grain prices. Since the budgets of food aid agencies are set well in advance, a rise in prices shrinks food assistance precisely when more help is needed. The U.N. World Food Programme, which supplies emergency food aid to more than 60 countries, has to cut shipments as prices soar. Meanwhile, over 7,000 children are dying each day from hunger and related illnesses.

When governments subsidize food-based biofuel production, they are in effect spending taxpayers' money to raise costs at the supermarket checkout counter. In the United States, the production of fuel ethanol was encouraged by a tax credit granted to fuel blenders for each gallon of ethanol they blended with gasoline. This tax credit expired at the end of 2011.

Still in place, however, is the Renewable Fuel Standard, which is seen by the U.S. Department of Agriculture as part of a strategy to "help recharge the rural American economy." This mandate requires that biofuel use ramp up to 36 billion gallons annually by 2022. Of this total, 16 billion gallons are slated to come from cellulosic feedstocks, such as cornstalks, grass, or wood chips.

Yet for the foreseeable future, production of those cellulose-based fuels has little chance of reaching such levels. Producing ethanol from sugars or starches like corn or sugarcane is a one-step process that converts the feedstock to ethanol. But producing ethanol from cellulosic materials is a two-step process: first the material must be broken down into sugar or starch, and then it is converted into ethanol. Furthermore, cellulosic feedstocks like corn stalks are much bulkier than feedstocks like corn kernels, so transporting them from distant fields to a distillery is much more costly. Removing agricultural residues such as corn stalks or wheat straw from the field to produce ethanol deprives the soil of needed organic matter.

The unfortunate reality is that the road to this ambitious cellulosic biofuel goal is littered with bankrupt firms that tried and failed to develop a process that would produce an economically viable fuel. Despite having the advantage of not being directly part of the food supply, cellulosic ethanol has strong intrinsic characteristics that put it at a basic disadvantage compared with grain ethanol, so it may never become economically viable.

The mandate from the European Union (EU) requiring that 10 percent of its transportation energy come from renewable sources, principally biofuels, by 2020 is similarly ambitious. Among international agribusiness firms, this is seen as a reason to acquire land, mostly in Africa, on which to produce fuel for export to Europe. Since Europe relies primarily on diesel fuel for its cars, the investors are looking at crops such as the oil palm and jatropha, a relatively low-yielding oil-bearing shrub, as a source of diesel fuel.

There is growing opposition to this EU goal from environmental groups, the European Environment Agency, and many other stakeholders. They object to the deforestation and the displacement of the poor that often results from such "land grabbing." (See Chapter 10.) They are also concerned that, by and large, biofuels do not deliver the promised climate benefits.

The biofuel industry and its proponents have argued that greenhouse gas emissions from biofuels are lower than those from gasoline, but this has been challenged by a number of scientific studies. Indeed, there is growing evidence that biofuel production may contribute to global warming rather than ameliorate it. A study led by Nobel prize–winning chemist Paul Crutzen at the Max Planck Institute for Chemistry in Germany reports that the nitrogen fertilizers used to produce biofuel crops release "nitrous oxide emissions large enough to cause climate warming instead of cooling."

A report from Rice University that carefully examined the greenhouse gas emissions question concluded that "it is uncertain whether existing biofuels production provides any beneficial improvement over traditional gasoline, after taking into account land use changes and emissions of nitrous oxide. Legislation giving biofuels preferences on the basis of greenhouse gas benefits should be avoided." The U.S. National Academy of Sciences also voiced concern about biofuel production's negative effects on soils, water, and the climate.

There is some good news on the issue of food or fuel. An April 2012 industry report notes that "the world ethanol engine continues to sputter." U.S. ethanol production likely peaked in 2011 and is projected to drop 2 percent in 2012. An even greater decline in U.S. ethanol production is likely in 2013 as oil prices weaken and as heat and drought in the U.S. Midwest drive corn prices upward. For many distillers, the profit margin disappeared in 2012. In early July 2012, Valero Energy Corporation, an oil company and a major ethanol producer, reported it was idling the second of its 10 ethanol distilleries. Numerous other distilleries are on the verge of shutting down.

If the ethanol mandate were phased out, U.S. distillers would have even less confidence in the future marketability of ethanol. In a world of widely fluctuating oil and grain prices, ethanol production would not always be profitable.

Beyond this, the use of automotive fuel in the United States, which peaked in 2007, fell 11 percent by 2012. Young people living in cities are simply not as car-oriented as their parents were. They are not part of the car culture. This helps explain why the size of the U.S. motor vehicle fleet, after climbing for a century, peaked at 250 million in 2008. It now appears that the fleet size will continue to shrink during this decade.

In addition, the introduction of more stringent U.S. auto fuel-efficiency standards means that gasoline use by new cars sold in 2025 will be half that of new cars sold in 2010. As older, less efficient cars are retired and fuel use declines, the demand for grain-based ethanol for blending will also decline.

Within the automobile sector, a major move to plug-in hybrids and all-electric cars will further reduce the use of gasoline. If this shift is accompanied by investment in thousands of wind farms to feed cheap electricity into the grid, then cars could run largely on electricity for the equivalent cost of 80¢ per gallon of gasoline.

There is also a growing public preference for walking, biking, and using public transportation wherever possible. This reduces not only the demand for cars and gasoline but also the paving of land for roads and parking lots.

Whether viewed from an environmental or an economic vantage point, we would all benefit by shifting from liquid fuels to electrically driven vehicles. Using electricity from wind farms, solar cells, or geothermal power plants to power cars will dramatically reduce carbon emissions. We now have both the electricity-generating technologies and the automotive technologies to create a clean, carbon-free transportation system, one that does not rely on either the use of oil or the conversion of food crops into fuel.

Data, endnotes, and additional resources can be found at Earth Policy Institute, www.earth-policy.org.

5

Eroding Soils Darkening Our Future

In 1938 Walter Lowdermilk, a senior official in the Soil Conservation Service of the U.S. Department of Agriculture, traveled abroad to look at lands that had been cultivated for thousands of years, seeking to learn how these older civilizations had coped with soil erosion. He found that some had managed their land well, maintaining its fertility over long stretches of history, and were thriving. Others had failed to do so and left only remnants of their illustrious pasts.

In a section of his report entitled "The Hundred Dead Cities," he describes a site in northern Syria, near Aleppo, where ancient buildings are still standing in stark isolated relief, but they are on bare rock. During the seventh century, the thriving region had been invaded, initially by a Persian army and later by nomads out of the Arabian Desert. In the process, soil and water conservation practices used for centuries were abandoned. Lowdermilk noted, "Here erosion had done its worst. If the soils had remained, even though the cities were destroyed and the populations dispersed, the area might be repeopled again and the cities rebuilt. But now that the soils are gone, all is gone."

The thin layer of topsoil that covers the earth's land

surface was formed over long stretches of geological time as new soil formation exceeded the natural rate of erosion. Sometime within the last century, soil erosion began to exceed new soil formation. Now, nearly a third of the world's cropland is losing topsoil faster than new soil is forming, reducing the land's inherent fertility. Soil that was formed on a geological time scale is being lost on a human time scale.

Scarcely six inches thick, this thin film of soil is the foundation of civilization. Geomorphologist David Montgomery, in *Dirt: The Erosion of Civilizations*, describes soil as "the skin of the earth—the frontier between geology and biology."

The erosion of soil by wind and water is a worldwide challenge. For the rangelands that support 3.4 billion head of cattle, sheep, and goats, the threat comes from the overgrazing that destroys vegetation, leaving the land vulnerable to erosion. Rangelands, located mostly in semiarid regions of the world, are particularly vulnerable to wind erosion.

In farming, erosion results from plowing land that is steeply sloping or too dry to support agriculture. Steeply sloping land that is not protected by terraces, perennial crops, strip cropping, or in some other way loses soil during heavy rains. Thus the land hunger that drives farmers up mountainsides fuels erosion.

In the United States, wind erosion is common in the semiarid Great Plains, where the country's wheat production is concentrated. In the U.S. Corn Belt, in contrast, where most of the country's corn and soybeans are grown, the principal threat to soil is water erosion. This is particularly true in the states with rolling land and plentiful rainfall, such as Iowa and Missouri.

Water erosion of soil has indirect negative effects, which can be seen in the silting of reservoirs and in muddy, silt-laden rivers flowing into the sea. Pakistan's two large reservoirs, Mangla and Tarbela, which store Indus River water for the country's vast irrigation network, have lost a third of their storage capacity over the last 40 years as they fill with silt from deforested watersheds.

Evidence of wind erosion is highly visible in the form of dust storms. When vegetation is removed either by overgrazing or overplowing, the wind begins to blow soil particles away, sometimes creating dust storms. Because the particles are small, they can remain airborne over great distances. Once they are largely gone, leaving mostly larger particles, sandstorms begin. These are local phenomena, often resulting in dune formation and the abandonment of both farming and grazing. The emergence of sandstorms marks the final phase in the desertification process.

The vast twentieth-century expansion in world food production pushed agriculture onto highly vulnerable land in many countries. The overplowing of the U.S. Great Plains during the late nineteenth and early twentieth centuries, for example, led to the 1930s Dust Bowl. This was a tragic era in U.S. history—one that forced hundreds of thousands of farm families to leave the Great Plains. Many migrated to California in search of a new life, a movement immortalized in John Steinbeck's *The Grapes of Wrath*.

Three decades later, history repeated itself in the Soviet Union. The Virgin Lands Project, a huge effort between 1954 and 1960 to convert grassland into grainland, led to the plowing of an area for grain that exceeded the current grainland in Canada and Australia combined. Initially this resulted in an impressive expansion in Soviet grain production, but the success was short-lived, as a dust bowl quickly developed there too.

Kazakhstan, at the center of the Virgin Lands Project, saw its grainland area peak at 25 million hectares in the early 1980s. After dropping to 11 million hectares in 1999, the area expanded again, reaching 17 million hectares in 2009, but then began once more to decline. Even on this reduced area, the average grain yield today is scarcely 1 ton per hectare—a far cry from the 7 tons per hectare that farmers get in France, Western Europe's leading wheat producer and exporter. The precipitous drop in Kazakhstan's grain area illustrates the price that countries pay for overplowing and overgrazing.

Today two giant new dust bowls have formed. One is centered in the Asian heartland in northwestern China and western Mongolia. The other is in the African Sahel—the savannah-like ecosystem that stretches across Africa from Somalia and Ethiopia in the east to Senegal and Mauritania in the west. It separates the Sahara Desert from the tropical rainforests to the south. Both of these newer dust bowls are massive in scale, dwarfing anything the world has seen before.

China may face the biggest challenge of all. After the economic reforms in 1978 that shifted the responsibility for farming from large state-organized production teams to individual farm families, China's cattle, sheep, and goat numbers spiraled upward. A classic tragedy of the commons was unfolding. The United States, a country with comparable grazing capacity, has 94 million cattle, a somewhat larger herd than China's 84 million. But when it comes to sheep and goats, the United States has a combined population of only 9 million, whereas China has 285 million. Concentrated in China's western and northern provinces, these animals are stripping the land of its protective vegetation. The wind then does the rest, removing the soil and converting rangeland into desert.

Wang Tao, one of the world's leading desert scholars, reports that from 1950 to 1975 an average of 600 square miles of land turned to desert each year. Between 1975 and 1987, this climbed to 810 square miles a year. From then until the century's end, it jumped to 1,390 square miles of land going to desert annually.

A U.S. Embassy report entitled "Desert Mergers and Acquisitions" describes satellite images showing two of China's largest deserts, the Badain Jaran and Tengger, expanding and merging to form a single, larger desert overlapping Inner Mongolia and Gansu Provinces. To the west in Xinjiang Province, two even larger deserts—the Taklimakan and Kumtag—are also heading for a merger. Highways running through the shrinking region between them are regularly inundated by sand dunes.

In some places, people become aware of soil erosion when they suffer through dust storms. On March 20, 2010, for example, a suffocating dust storm enveloped Beijing. The city's weather bureau took the unusual step of describing the air quality as hazardous, urging people to stay inside or to cover their faces if they were outdoors. Visibility was low, forcing motorists to drive with their lights on in daytime.

Beijing was not the only area affected. This particular dust storm engulfed scores of cities in five provinces, directly affecting over 250 million people. Nor was it an isolated incident. Every spring, residents of eastern Chinese cities, including Beijing and Tianjin, hunker down as the dust storms begin. Along with having difficulty breathing and dealing with dust that stings the eyes, people must constantly struggle to keep dust out of their homes and to clear doorways and sidewalks of dust and sand. Farmers and herders whose livelihoods are blowing away are paying an even higher price.

These huge dust storms originating in northwestern and north central China and western Mongolia form in the late winter and early spring. On average more than 10 major dust storms leave this region and move across the country's heavily populated northeast each year. These dust storms affect not only China but neighboring countries as well. The March 2010 dust storm arrived in South Korea soon after leaving Beijing. It was described by the Korean Meteorological Administration as the worst dust storm on record.

Highly detailed media accounts of these storms are not always readily available, but Howard French described in the *New York Times* a Chinese dust storm that had reached South Korea on April 12, 2002. The country, he said, was engulfed by so much dust from China that people in Seoul were literally gasping for breath. Schools were closed, airline flights were cancelled, and clinics were overrun with patients who were having trouble breathing. Retail sales fell. Koreans have come to dread the arrival of what they call "the fifth season"—the dust storms of late winter and early spring.

The situation continues to deteriorate. Korea's Ministry of Environment reports that the country suffered dust storms on average for 39 days in the 1980s, 77 days in the 1990s, and 118 days from 2000 to 2011. These data suggest that the degradation of land is accelerating. Unfortunately, there is nothing in prospect to arrest and reverse this trend.

While people living in China and South Korea are all too familiar with dust storms, the rest of the world typically only learns about this fast-growing ecological catastrophe when the massive soil-laden storms leave that region. On April 18, 2001, for instance, the western United States—from the Arizona border north to Canada—was blanketed with dust. It came from a huge dust storm that originated in northwestern China and Mongolia on April 5th.

Another consequence of dust storms is the economic disruption that they cause in cities, whether it is Beijing or any of dozens of other cities in northeastern China or

South Korea. Dust storms can disrupt business, reduce retail sales, close schools, and even temporarily close governments in some cases. Each of these disruptions brings its own cost. Sometimes the effects are remote from the site of the dust, as when dust particles from African storms damage coral reefs in the Caribbean, adversely affect fishing and tourism.

Africa is suffering heavy losses of soil from wind erosion. Andrew Goudie, Emeritus Professor in Geography at Oxford University, reports that dust storms originating over the Sahara—once rare—are now commonplace. He estimates they have increased tenfold during the last half-century. Among the countries most affected by topsoil loss via dust storms are Niger, Chad, northern Nigeria, and Burkina Faso. In Mauritania, in Africa's far west, the number of dust storms jumped from 2 a year in the early 1960s to 80 in 2004.

The Bodélé Depression, a vast low-lying region in northeastern Chad, is the source of an estimated 1.3 billion tons of dust a year, up tenfold from 1947, when measurements began. Dust storms leaving Africa typically travel west across the Atlantic, depositing dust in the Caribbean. The 2–3 billion tons of fine soil particles that leave Africa each year in dust storms are slowly draining the continent of its fertility and hence its biological productivity.

Nigeria, Africa's most populous country, is losing 868,000 acres of rangeland and cropland to desertification each year. The government considers the loss of productive land to desert to be far and away its leading environmental problem. No other environmental change threatens to undermine its economic future so directly. Conditions will only get worse if Nigeria continues on its current population trajectory toward 390 million people by 2050.

While Nigeria's human population has increased from 47 million in 1961 to 167 million in 2012, nearly a fourfold expansion, its population of livestock has grown from roughly 8 million to 109 million head. With the forage needs of Nigeria's 17 million head of cattle and 92 million sheep and goats exceeding the sustainable yield of the country's grasslands, the country is slowly turning to desert. (See Figure 5–1.)

In fact, Nigeria presents a textbook case of how mounting human and livestock population pressures reduce vegetative cover. Most notably, growth in the goat population relative to sheep and cattle is a telltale indicator of grassland ecosystem deterioration. As grasslands deteriorate from overgrazing, grass is typically replaced by desert shrubs. In such a degraded environment as Nigeria's, sheep and cattle do not fare well, but goats—being particularly hardy ruminants—forage on the shrubs.

Between 1970 and 2010, the world cattle population increased by 32 percent, the sheep population was unchanged, but the goat population more than doubled. This dramatic shift in the composition of the livestock

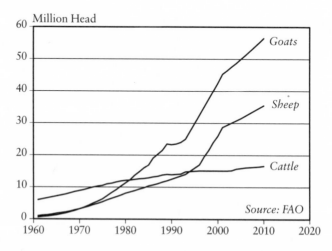

Figure 5–1. *Grazing Livestock in Nigeria, 1961–2010*

herd, with goats now in such a dominant role, promises continuing grassland deterioration and accelerating soil erosion.

Growth in the goat population has been dramatic in some other developing countries as well, particularly in Africa and Asia, which combined account for 90 percent of the world's goats. While Pakistan's cattle population more than doubled between 1961 and 2010, and the sheep population nearly tripled, the goat population grew almost sevenfold. In Bangladesh, cattle and sheep populations have grown only modestly since 1980, while the population of goats has quadrupled. In 1985, Mali had roughly equal populations of cattle, sheep, and goats, but while its cattle and sheep populations have remained relatively stable since then, its goat population has more than tripled.

Meanwhile, on the northern fringe of the Sahara, countries such as Algeria and Morocco are attempting to halt the desertification that is threatening their fertile croplands. Algerian president Abdelaziz Bouteflika says that Algeria is losing 100,000 acres of its most fertile lands to desertification each year. For a country that has only 7.7 million acres of grainland, this is not a trivial loss. Among other measures, Algeria is planting its southernmost cropland in perennials, such as fruit orchards, olive orchards, and vineyards—crops that can help keep the soil in place.

India is also in a war with expanding deserts. With scarcely 2 percent of the world's land area, India is struggling to support 18 percent of the world's people and 15 percent of its cattle. According to a team of scientists at the Indian Space Research Organization, 25 percent of India's land surface is slowly turning into desert. It thus comes as no surprise that many of India's cattle are emaciated.

In Afghanistan, a U.N. Environment Programme (UNEP) team reports that in the Sistan region in the country's southwest "up to 100 villages have been submerged by windblown dust and sand." The Registan Desert is migrating westward, encroaching on agricultural areas. In the country's northwest, sand dunes are moving onto agricultural land in the upper Amu Darya basin, their path cleared by the loss of stabilizing vegetation due to firewood gathering and overgrazing. The UNEP team observed sand dunes as high as a five-story building blocking roads, forcing residents to establish new routes.

An Afghan Ministry of Agriculture and Food report sounds the alarm: "Soil fertility is declining,...water tables have dramatically fallen, de-vegetation is extensive and soil erosion by water and wind is widespread." After three decades of armed conflict and the related deprivation and devastation, Afghanistan's forests are nearly gone. Seven southern provinces are losing cropland to encroaching sand dunes. And like many failing states, even if Afghanistan had appropriate environmental policies, it lacks the law enforcement capacity to implement them.

Iraq, suffering from nearly a decade of war and recent drought and chronic overgrazing and overplowing, is now losing irrigation water to its upstream riparian neighbor—Turkey. The reduced river flow—combined with the deterioration of irrigation infrastructure, the depletion of aquifers, the shrinking irrigated area, and the drying up of marshlands—is drying out Iraq. The Fertile Crescent, the cradle of civilization, may be turning into a dust bowl.

Dust storms are forming with increasing frequency in western Syria and northern Iraq. In July 2009 a dust storm raged for several days in what was described as the worst such storm in Iraq's history. As it traveled eastward into Iran, the authorities in Tehran closed government offices, private offices, schools, and factories. Although this new dust bowl is small compared with those centered in northwest China and across central Africa, it is nonetheless an unsettling new development in this region.

Iran—with 76 million people—illustrates the pressures facing the Middle East. With 9 million cattle and 80 million sheep and goats—the source of wool for its fabled rug-making industry—Iran's rangelands are deteriorating from overstocking. Mohammad Jarian, who heads Iran's Anti-Desertification Organization, reported in 2002 that sandstorms had buried 124 villages in the southeastern province of Sistan-Balochistan, forcing their abandonment. Drifting sands had covered grazing areas, starving livestock and depriving villagers of their livelihoods.

As countries lose their topsoil, they eventually lose the capacity to feed themselves. Among those facing this problem are Lesotho, Mongolia, North Korea, and Haiti. Lesotho, one of Africa's smallest countries, with only 2 million people, is paying a heavy price for its soil losses. A U.N. team visited in 2002 to assess its food prospects. Their finding was straightforward: "Agriculture in Lesotho faces a catastrophic future; crop production is declining and could cease altogether over large tracts of the country if steps are not taken to reverse soil erosion, degradation, and the decline in soil fertility."

Michael Grunwald reported in the *Washington Post* that nearly half of the children under five in Lesotho are stunted physically. "Many," he wrote, "are too weak to walk to school." Over the last decade, Lesotho's grain harvest dropped by half as its soil fertility fell. Its collapsing agriculture has left the country heavily dependent on food imports.

A similar situation exists in Mongolia, where over the last 20 years more than half of the wheatland has been abandoned and wheat yields have started to fall, shrinking its harvest. Mongolia now imports nearly 20 percent of its wheat. At the same time, North Korea, largely deforested and suffering from flood-induced soil erosion and land degradation, has watched its yearly grain harvest fall from a peak of almost 6 million tons during the 1980s to scarcely 3 million tons per year today.

In the western hemisphere, Haiti—one of the early failing states—was largely self-sufficient in grain 40 years ago. Since then it has lost nearly all its forests and much of its topsoil, forcing it to import over half of its grain. It is now heavily dependent on U.N. World Food Programme lifelines.

The accelerating loss of topsoil is slowly but surely reducing the earth's inherent biological productivity. The shrinking area of productive land and the earth's steadily expanding human population are on a collision course. Soil erosion and land degradation issues are local, but their effect on food security is global.

Data, endnotes, and additional resources can be found at Earth Policy Institute, www.earth-policy.org.

6

Peak Water and Food Scarcity

Although many analysts are concerned about the depletion of oil resources, the depletion of underground water resources poses a far greater threat to our future. While there are substitutes for oil, there are none for water. Indeed, modern humans lived a long time without oil, but we would live for only a matter of days without water.

Not only are there no substitutes for water, but the world needs vast amounts of it to produce food. As adults, each of us drinks nearly 4 liters of water a day in one form or another. But it takes 2,000 liters of water—500 times as much—to produce the food we consume each day.

Since food is such an extraordinarily water-intensive product, it comes as no surprise that 70 percent of world water use is for irrigation. Although it is now widely accepted that the world is facing severe water shortages, not everyone realizes that a future of water shortages will also be a future of food shortages.

The use of irrigation to expand food production goes back some 6,000 years. Indeed, the development of irrigation using water from the Tigris and Euphrates Rivers set the stage for the emergence of the Sumerian civilization, and it was the Nile that gave birth to ancient Egypt.

Throughout most of history, irrigation spread rather slowly. But in the latter half of the twentieth century it underwent a rapid expansion. In 1950, there were some 250 million acres of irrigated land in the world. By 2000, the figure had nearly tripled to roughly 700 million acres. After these several decades of rapid increase, however, the growth in irrigated area has slowed dramatically since the turn of the century, expanding only 9 percent from 2000 to 2009. Given that governments are much more likely to report increases than decreases, the recent net growth in irrigated area may be even smaller. This dramatic loss of momentum in irrigation expansion, coupled with the aquifer depletion that is already reducing irrigated area in some countries, suggests that peak water may now be on our doorstep.

The trend in irrigated land area per person is even less promising. For the last half-century, the irrigated area has been expanding—but not as fast as population. As a result, the irrigated area per person today is 10 percent less than it was in 1960. With so many aquifers being depleted and more and more irrigation wells going dry, this shrinkage in irrigated area per person is likely not only to continue but to accelerate in the years ahead.

Roughly 40 percent of the world grain harvest is grown on irrigated land. The rest is rainfed. Among the big three grain producers—China, India, and the United States— the role of irrigation varies widely. In China, four fifths of the grain harvest comes from irrigated land. For India it is three fifths, and for the United States, only one fifth. Asia, where rice is the staple food, totally dominates the world irrigated area.

Farmers use both surface and underground water for irrigation. Surface water is typically stored behind dams on rivers and then channeled onto the land through a network of irrigation canals. Historically, and notably from 1950 until 1975, when most of the world's large dams were built, this was the main source of growth in world irrigated area. During the 1970s, however, as the sites for new dams diminished, attention shifted from building dams to drilling wells for access to underground water.

Most underground water comes from aquifers that are regularly replenished with rainfall; these can be pumped indefinitely as long as water extraction does not exceed recharge. A small minority of aquifers are fossil aquifers, however, containing water put there eons ago. Since these do not recharge, irrigation ends once they are pumped dry. Among the more prominent fossil aquifers are the Ogallala underlying the U.S. Great Plains, the deep aquifer under the North China Plain, and the Saudi aquifers.

Given a choice, farmers generally prefer having their own wells because it enables them to control the timing and amount of water delivered with a precision that is not possible with large, centrally managed canal irrigation systems. Pumps let them apply water precisely when the crop needs it, thus achieving higher yields than with large-scale, river-based irrigation systems. Forty percent of world irrigated area is now dependent on underground water. As world demand for grain has climbed, farmers have drilled more and more irrigation wells with little concern for how many the local aquifers could support. As a result, water tables are falling and millions of irrigation wells are either going dry or are on the verge of doing so.

As groundwater use for irrigation expands, so does the grain harvest. But if the pumping surpasses the sustainable yield of the aquifer, aquifers are depleted. When this happens, the rate of irrigation pumping is necessarily reduced to the aquifer's natural rate of recharge. At this point, grain production declines too.

The resulting water-based "food bubbles," which create a short-term false sense of security, can now be found in some 18 countries that contain more than half the world's people. In these countries, food is being produced by drawing down water reserves. This group includes China, India, and the United States. (See Table 6–1.)

In Saudi Arabia, pumping is fast depleting the country's major aquifers. After the Arab oil-export embargo in the 1970s, the Saudis realized that since they were heavily dependent on imported grain they were vulnerable to a grain counter-embargo. Using oil-drilling technology, they tapped into aquifers far below the desert to produce irrigated wheat. In a matter of years, the kingdom was self-sufficient in wheat, a food staple.

But after more than 20 years of wheat self-sufficiency, the Saudis announced in January 2008 that their aquifers were largely depleted and they would be phasing out wheat production. Between 2007 and 2011, the wheat harvest of just under 3 million tons dropped by nearly half. At this rate the Saudis likely will harvest their last wheat crop by 2016, as planned, and will then be totally dependent on imported grain to feed nearly 30 million people.

The unusually rapid phaseout of wheat farming in Saudi Arabia is due to two factors. First, in this arid country there is little farming without irrigation. Second, its irrigation depends almost entirely on fossil aquifers. The desalted seawater that Saudi Arabia uses in its cities is far too costly for large-scale irrigation use.

Saudi Arabia's growing food insecurity has even led it to buy or lease land in several other countries, importantly Ethiopia and Sudan. (See Chapter 10.) The Saudis are planning to produce food for themselves with the land and water resources of other countries to augment their fast-growing grain purchases in the world market.

In neighboring Yemen, replenishable aquifers are also

Table 6–1. *Countries Overpumping Aquifers in 2012*

Country	Population (million)
Afghanistan	33
China	1,354
India	1,258
Iran	76
Iraq	34
Israel	8
Jordan	6
Lebanon	4
Mexico	116
Morocco	33
Pakistan	180
Saudi Arabia	29
South Korea	49
Spain	47
Syria	21
Tunisia	11
United States	316
Yemen	26
Total	3,599

Source: Earth Policy Institute, with populations from U.N. Population Division.

being pumped well beyond the rate of recharge, and the deeper fossil aquifers are being rapidly depleted too. As a result, water tables are falling throughout Yemen by some 2 meters per year. In the capital, Sana'a—home to 2 million people—a 2006 report indicated that tap water was available only once every 4 days; in Taiz, a smaller city to the south, it was once every 20 days.

Yemen, where population growth is spiraling out of control, is fast becoming a hydrological basket case. With water tables falling, the grain harvest has shrunk by one half over the last 40 years, while demand has continued its steady rise. As a result, the Yemenis now import more than 80 percent of their grain. With its meager oil exports falling, with no industry to speak of, and with nearly 60 percent of its children physically stunted and chronically undernourished, this poorest of the Middle East Arab countries is facing a bleak and turbulent future.

The likely result of the depletion of Yemen's aquifers, which will lead to further shrinkage of its harvest and spreading hunger, is social collapse. Already a failing state, it may well devolve into a group of tribal fiefdoms, warring over whatever meager water resources remain. For the international community, the risk is that Yemen's internal conflicts could spill over its lengthy, unguarded border with Saudi Arabia.

In addition to the bursting food bubble in Saudi Arabia and the fast-deteriorating water situation in Yemen, two other populous countries in the region—Syria and Iraq—have water troubles. Some of these arise from the reduced flows of the Euphrates and Tigris Rivers, which both countries depend on for irrigation water. Turkey, which controls the headwaters of both these rivers, is in the midst of a massive dam building program that is slowly reducing downstream flows. Although all three countries have discussed water-sharing arrangements, Turkey's ambitious plans to expand both its hydropower generation and irrigated area are being fulfilled partly at the expense of its downstream neighbors.

This is nowhere more evident than in Turkey's massive diversion of water from the Euphrates River by its large southeast Anatolia project. Harald Frederiksen, one of the World Bank's leading water management consultants, says that Turkey's retention of Euphrates and Tigris River flows has "severely reduced the millennia-old supply to the other riparians." Some analysts estimate that Syria will lose at least 30 percent of its water supply and Iraq, the last country in the Tigris-Euphrates flow, at least 60 percent. Others, who see an even grimmer water future in the region, believe Syria could lose 50 percent and Iraq up to 90 percent. With the loss of irrigation water, many Iraqis are abandoning their land and migrating to cities. Frederiksen notes, "The lower riparians' desperate situation today presents the world community with a highly volatile international security situation."

Given the uncertainty of river water supplies, farmers in Syria and Iraq have drilled many wells for irrigation, leading to overpumping and falling water tables in both countries. With wells going dry, Syria's grain harvest has fallen by one third since peaking at roughly 7 million tons in 2001. In Iraq, the grain harvest has fallen by one sixth since peaking at 4.5 million tons in 2002.

Jordan, with over 6 million people, is also on the ropes agriculturally, due to unsustainable aquifer withdrawals. The Ministry of Water and Irrigation estimates that groundwater withdrawals are nearly twice the sustainable yield, causing the overexploitation and abandonment of both municipal and irrigation wells. Forty or so years ago, the country was producing over 300,000 tons of grain per year. Today, it produces only 55,000 tons and must import over 90 percent of the grain it consumes. In the region, only Lebanon has managed to avoid a decline in grain production.

Thus in the Arab Middle East, where populations are growing fast, the world is seeing the first regional collision between population growth and water supply. For the first time in history, water shortages are shrinking the grain harvest in an entire geographic region—with nothing in sight to arrest the decline. Because of the failure of governments in the region to mesh population and water policies, each day now brings 9,000 more people to feed and less irrigation water with which to feed them.

A similar prospect of spreading water shortages threatens China. Although surface water is widely used for irrigation, the principal concern is the groundwater situation in the northern half of the country, where rainfall is low and water tables are falling everywhere. This includes the highly productive North China Plain, which stretches from north of Beijing south toward Shanghai and produces half of the country's wheat and a third of its corn.

The scale of overpumping in the North China Plain suggests that some 130 million Chinese are being fed with grain produced with the unsustainable use of water. Farmers in this region are pumping from two aquifers: the so-called shallow aquifer, which is rechargeable but largely depleted, and the deep fossil aquifer. Once the latter is depleted, the irrigated agriculture dependent on it will end, forcing farmers back to rainfed farming.

China has had ample warning. A groundwater survey done more than a decade ago by the Geological Environment Monitoring Institute (GEMI) in Beijing found that under Hebei Province, in the heart of the North China Plain, the average level of the deep aquifer dropped 2.9 meters (nearly 10 feet) in 2000. Around some cities in the province, it fell by 6 meters in that one year alone. He Qingcheng, director of the GEMI groundwater monitoring team, notes that as the deep aquifer under the North China Plain is depleted, the region is losing its last water reserve—its only safety cushion.

In a 2010 interview with *Washington Post* reporter Steven Mufson, He Qingcheng noted that Beijing was drilling down 1,000 feet to reach water—five times deeper than 20 years ago. His concerns are mirrored in the unusually strong language of a World Bank report on China's water situation that foresees "catastrophic consequences for future generations" unless water use and supply can quickly be brought back into balance.

The problem may be even more serious in India, simply because the margin between actual food consumption and survival is so thin. In this global epicenter of well drilling, where farmers have drilled 21 million irrigation wells, water tables are dropping in much of the country. Among the states most affected are Punjab, Haryana, Rajasthan, and Gujarat in the north and Tamil Nadu in the south. The wells, powered by heavily subsidized electricity, are dropping water tables at an accelerating rate. In North Gujarat, the water table is falling by 6 meters, or 20 feet, per year. In some states, half of all electricity is now used to pump water.

In Tamil Nadu, a state of 72 million people, falling water tables are drying up wells. Kuppannan Palanisami of Tamil Nadu Agricultural University says that falling water tables have dried up 95 percent of the wells owned by small farmers, reducing the irrigated area in the state by half over the last decade.

As water tables fall, small farmers often lose out because they lack the capital required to drill deeper. Larger farmers in India are using modified oil-drilling technology to reach water, going as deep as 1,000 feet in some locations. Pumping from such depths is energy-intensive and costly. In communities where underground water sources have dried up entirely, all agriculture is rainfed and drinking water is trucked in. Tushaar Shah, a senior fellow at the International Water Management Institute, says, "When the balloon bursts, untold anarchy will be the lot of rural India."

The United States is also depleting its aquifers. In most of the leading U.S. irrigation states, the irrigated area has peaked and begun to decline. In California, historically the irrigation leader, a combination of aquifer depletion and the diversion of water to fast-growing cities has reduced irrigated area from nearly 9 million acres in 1997 to 8 million acres in 2007. In Texas, the irrigated area peaked in 1978 at 7 million acres, falling to some 5 million acres in 2007 as the thin southern end of the Ogallala aquifer that underlies much of the Texas panhandle was depleted.

Other states with shrinking irrigated area include Arizona, Colorado, and Florida. Colorado has watched its irrigated area shrink for the last few decades. Researchers there project a loss of up to 700,000 acres of irrigated land between 2010 and 2050, which is roughly one fifth of the state's total. All three states are suffering from both aquifer depletion and the diversion of water to urban centers. And now that the growth in irrigated area in the states where it has rapidly expanded over the last decade or so, such as Nebraska and Arkansas, is starting to level off, the prospects for any national growth in irrigated area have faded. With water tables falling as aquifers are depleted under the Great Plains and California's Central Valley, and with fast-growing cities in the Southwest taking more and more water, U.S. irrigated area appears to have peaked and begun a long-term decline.

In Mexico, a largely semiarid country that is home to 116 million people, the demand for water is outstripping supply. Mexico City's water problems are well known, but rural areas are also suffering. In the agricultural state of Guanajuato, the water table is falling by 6 feet or more a year. In the northwestern wheat-growing state of Sonora, farmers once pumped water from the Hermosillo aquifer at a depth of 40 feet. Today, they pump from over 400 feet. With 58 percent of all water extraction in Mexico coming from aquifers that are being overpumped, Mexico's food bubble may burst soon.

In many of the world's river basins, tensions are build-ing as competition for scarce water intensifies. Egypt, at the lower reaches of the Nile River, with a population of 84 million people in a country where it rarely rains, is highly vulnerable. Egypt either imports its wheat or imports the water to produce it via the Nile River. And since Egypt is a nation of bread eaters, what happens to its wheat supply is a matter of intense public interest.

The Nile Waters Agreement, which Egypt and Sudan signed in 1959, allocated 75 percent of the river's flow to Egypt, 25 percent to Sudan, and none to Ethiopia. Howev-er, this agreement has largely become void in practice, in the face of wealthy foreign governments and international agribusiness firms who are snatching up large swaths of arable land in the upper Nile basin. While these deals are typically described as land acquisitions, they are also, in effect, water acquisitions.

Unfortunately for Egypt, both Ethiopia and the two Sudans—the upstream countries that together occupy three fourths of the Nile River basin—are among the principal targets of land acquisitions. In South Sudan, a full 4 percent of the country's land area had already been acquired by foreign investors when it achieved indepen-dence. Demands for water in the Nile basin are such that there is little of the river left when it eventually reaches the Mediterranean.

When competing for Nile water, Cairo now must deal with a number of governments and commercial interests that were not party to the 1959 agreement. Moreover, Ethiopia has announced plans to build a huge hydroelec-tric dam on its branch of the Nile, which would reduce the water flow to Egypt even more.

Because Egypt's wheat yields are already among the world's highest, it has little potential to raise its land productivity further. With its population projected to reach 101 million by 2025, finding enough food and water is an imminent and daunting challenge.

Egypt's plight could become part of a larger, more troubling scenario. Its upstream Nile neighbors—Sudan and South Sudan, with 46 million people, and Ethiopia, with 87 million—are growing even faster, increasing the need for water to produce food. Projections by the United Nations show the combined population of these four Nile-basin countries increasing from 216 million at present to 272 million by 2025.

The Nile is not the only river whose waters are fully allocated. In the southwestern United States, the Colorado River originates in the Rocky Mountains of Colorado and flows to the southwest, theoretically entering the Gulf of California. But now in fact it rarely reaches the Gulf. It is the principal source of irrigation water in the southwestern part of the United States, supplying water to Colorado, Utah, Nevada, Arizona, and California. Major cities such as Phoenix, San Diego, and Los Angeles also depend on its water.

A similar situation is unfolding in the Mekong River basin. China, which controls the headwaters of the Mekong, is building a number of dams, many of them for power generation. Although the water flows through these dams, each dam and the reservoir behind it reduces the amount of water reaching the countries in the lower part of the basin, such as Viet Nam, Thailand, Cambodia, and Laos, simply because of the evaporation factor. The rule of thumb for reservoirs is that each year 10 percent of the water they store evaporates. This loss of Mekong flow plus that from diversion in China threaten the downstream ecosystems, reducing fish populations and depriving many river dwellers of their livelihoods.

Another major river with a potential source of conflict is the Indus. Though a large part of the Indus water flow originates in India, most of the water is actually used in

Pakistan because of geography and the 1960 Indus Water Treaty. The Indus, flowing west from the Himalayas to the Indian Ocean, supplies water not only for Pakistan's Indus basin irrigation system, the world's largest, but also for the country's other needs. For much of the year, like the Colorado River, it now barely reaches the ocean.

Pakistan, with a population of 180 million people that is projected to reach 275 million by 2050, is facing trouble. Water expert John Briscoe writes in a World Bank study, "Pakistan is already one of the most water-stressed countries in the world, a situation which is going to degrade into outright water scarcity due to high population growth." He then notes that "the survival of a modern and growing Pakistan is threatened by water."

At the international level, water conflicts among countries dominate the headlines. But within countries it is the competition for water between cities and farms that preoccupies political leaders. Neither economics nor politics favors farmers. They almost always lose out to cities.

Indeed, in many countries farmers now face not only a shrinking water supply but also a shrinking share of that shrinking supply. In large areas of the United States, such as the southern Great Plains and the Southwest, virtually all water is now spoken for. The growing water needs of major cities and thousands of small towns often can be satisfied only by taking water from agriculture. As the value of water rises, more farmers are selling their irrigation rights to cities, letting their land dry up.

In the western United States, hardly a day goes by without the announcement of a new sale. Half or more of all sales are by individual farmers or their irrigation districts to cities and municipalities. Felicity Barringer, writing in the *New York Times* from California's Imperial Valley, notes that many fear that "a century after Colorado River water allowed this land to be a cornucopia, unfettered urban water transfers could turn it back into a desert."

Colorado, with a fast-growing population, has one of the world's most active water markets. Cities and towns of all sizes are buying irrigation water rights from farmers and ranchers. In the Arkansas River basin, which occupies the southeastern quarter of the state, Colorado Springs and Aurora (a suburb of Denver) have already bought water rights to one third of the basin's farmland. Aurora has purchased rights to water that was once used to irrigate 19,000 acres of cropland in the Arkansas valley. The U.S. Geological Survey estimates that 400,000 acres of farmland dried up statewide between 2000 and 2005.

Colorado is not alone in losing irrigation water. Farmers in India are also losing water to cities. This is strikingly evident in Chennai (formerly Madras), a city of 9 million on the east coast. As a result of the city government's inability to supply water to many of its residents, a thriving tank-truck industry has emerged that buys water from nearby farmers and hauls it to the city's thirsty residents.

For farmers near the city, the market price of water far exceeds the value of the crops they can produce with it. Unfortunately, the 13,000 tankers hauling water to Chennai are mining the region's underground water resources. Water tables are falling and shallow wells have gone dry. Eventually even the deeper wells will go dry, depriving rural communities of both their food supply and their livelihood. The intensifying competition for water at the local level led India's Minister of Water Resources to quip that he is actually the Minister of Water Conflicts.

In the competition for water between farmers on the one hand and cities and industries on the other, the economics do not favor agriculture. In countries such as China, where industrial development and the jobs associated with it are an overriding national economic goal, agriculture is becoming the residual claimant on the water supply.

In countries where virtually all water has been claimed, as in North Africa and the Middle East, cities can typically get more water only by taking it from irrigation. Countries then import grain to offset the loss of grain production. Since it takes 1,000 tons of water to produce 1 ton of grain, importing grain is the most efficient way to import water. Similarly, trading in grain futures is, in a sense, trading in water futures. To the extent that there is a world water market, it is embodied in the world grain market.

We live in a world where more than half the people live in countries with food bubbles based on overpumping. The question for each of these countries is not whether its bubble will burst, but when. And how will the government cope with it? Will governments be able to import grain to offset production losses? For some countries, the bursting of the bubble may well be catastrophic. For the world as a whole, the near-simultaneous bursting of several national food bubbles as aquifers are depleted could create unmanageable food shortages.

Given the sheer geographic scale of overpumping, the simultaneous fall of water tables among countries, and the accelerating rate of their drop, the need to stabilize water tables is urgent. Although falling water tables are historically recent, they now threaten the security of water supplies and, hence, of food supplies not only in the countries where this is occurring but throughout the world.

The gap between rising water use and the sustainable yield of aquifers grows larger each year, which means the drop in water tables each year is greater than the year before. Underlying the urgency of dealing with the fast-tightening water situation is the sobering realization that not a single country has succeeded in arresting the fall in its water tables. The fast-unfolding water crunch has not yet translated into food shortages at the global level. But if it is not addressed, it may do so soon.

7

Grain Yields Starting to Plateau

From the beginning of agriculture until the mid-twentieth century, growth in the world grain harvest came almost entirely from expanding the cultivated area. Rises in land productivity were too slow to be visible within a single generation. It is only within the last 60 years or so that rising yields have replaced area expansion as the principal source of growth in world grain production.

The transition was dramatic. Between 1950 and 1973 the world's farmers doubled the grain harvest, nearly all of it from raising yields. Stated otherwise, expansion during these 23 years equaled the growth in output from the beginning of agriculture until 1950. The keys to this phenomenal expansion were fertilization, irrigation, and higher-yielding varieties, coupled with strong economic incentives for production.

The first country to achieve a steady, sustained rise in grain yields was Japan, where the yield takeoff began in the 1880s. But for a half-century or so, it was virtually alone. Not until the mid-twentieth century did the United States and Western Europe launch a steady rise in grain yields. Shortly thereafter many other countries succeeded in boosting grain yields,

The average world grain yield in 1950 was 1.1 tons per

hectare. In 2011, it was 3.3 tons per hectare—a tripling of the 1950 level. Some countries, including the United States and China, managed to quadruple grain yields, and all within a human life span.

Some of the factors influencing grain yields are natural, while others are of human origin. Natural conditions of inherent soil fertility, rainfall, day length, and solar intensity strongly influence crop yield potentials. Several areas of cropland with inherently high fertility are found widely scattered around the world: in the U.S. Midwest (often called the Corn Belt), Western Europe, the Gangetic Plain of India, and the North China Plain. It is the incredibly deep and rich soils of the U.S. Midwest that enables the United States to produce 40 percent of the world corn crop and 35 percent of the soybean crop. The state of Iowa, for instance, produces more grain than Canada and more soybeans than China.

The area west of the Alps, stretching across France to the English Channel and up to the North Sea, is also naturally very productive land, enabling densely populated Western Europe to produce an exportable surplus of wheat.

The region in northern India spanning the Punjab and the Gangetic Plain is India's breadbasket. And the North China Plain produces half of China's wheat and a third of its corn.

Aside from inherent soil fertility, the level and timing of rainfall, which vary widely among geographic regions, also strongly influence the productivity of land. Much of the world's wheat, which is drought-tolerant, is grown without irrigation in regions with relatively low rainfall. Most wheat in the United States, Canada, and Russia, for example, is grown under these dryland conditions. Wheat is often grown in areas too dry or too cold to grow corn or rice.

Another natural factor that plays a major role in crop yields is day length. One reason that the United Kingdom and Germany have such high wheat yields is because they have a mild climate, compliments of the Gulf Stream, and can grow winter wheat. This wheat, planted in the fall, reaches several inches in height and then goes dormant as temperatures drop. With the arrival of spring, it grows rapidly, maturing during the longest days of the year in a high-latitude region that has very long days in May, June, and July. Wheat yields in these two northerly countries are close to 8 tons per hectare, somewhat higher than the 7 tons in France, simply because they are at a slightly higher latitude and thus have longer summer days.

The big differences between the United States and Western Europe are soil moisture and day length. In the United States, most wheat grows in the semiarid Great Plains, whereas in Europe it is produced on the well-watered, rainfed wheat fields of France, Germany, and the United Kingdom. The average U.S. wheat yield is scarcely 3 tons per hectare. But in Western Europe, wheat yields can range from 6 to 8 tons per hectare.

Just as long days promote high yields, the short days closer to the equator lead to relatively low yields. The advantage of the subtropical regions, however, is that they allow more than one crop per year, assuming sufficient soil moisture in the dry season. In land-scarce southern China, India, and other tropical/subtropical countries in Asia, double- or triple-cropping of rice is not uncommon. So although the yield per harvest is lower, the yield per year is much higher.

In northern India, for example, winter wheat with a summer rice crop is the dominant high-yielding combination. In China, combining winter wheat with corn as the summer crop in an annual cycle, plus the double-cropping of rice, enables the country to produce the world's largest grain harvest on a relatively modest area of arable land.

Solar intensity also plays an important role in determining the upper limits of crop yields. Rice yields in Japan, among the highest in Asia, are well below those in California. This is not because California's rice farmers are more skilled than their Japanese counterparts but because Japan's rice harvest grows mostly during the monsoon season, when there is extensive cloud cover, while California's rice fields bask in bright sunlight.

Within this framework of natural conditions that help determine yields, plant breeders have made impressive progress in exploiting the yield potential. Japan has been a long-time leader. The originally domesticated wheats and rices tended to be taller, enabling them to compete with weeds for sunlight. But with weed control either by hand or mechanical cultivation, Japanese plant breeders realized that the tall grain could be shortened. By shortening the straw, a greater share of the plant's photosynthate could be diverted to forming seeds, the edible part.

After Japanese "dwarf" wheats were introduced into the northwestern United States, Norman Borlaug, an agronomist based in Mexico, obtained some of the seeds in the early 1950s. He then introduced these dwarf wheats into other countries, including India and Pakistan, for testing under local growing conditions. Almost everywhere they were introduced they would double or even triple the yields of those from traditional wheat varieties. In Mexico, the dwarf wheats led to a quantum jump in wheat yields, nearly fourfold from 1950 to 2011.

Given the dramatic advances for the early dwarf wheats, in 1960 a similar effort with rice was launched at the newly created International Rice Research Institute (IRRI) at Los Baños in the Philippines. Under the leadership of Robert Chandler, scientists there drew on the experience with wheat to come up with some high-yielding dwarf rice varieties that were, like the wheats, widely adopted. IR8, one of the early strains, easily doubled yields in many countries. It was the first of many new highly productive rice strains to come from IRRI.

The new dwarf wheats and rices had the genetic potential to respond well to both irrigation and fertilizer. When fertilizer was applied to the old tall-strawed varieties, the plant would often fall over in a storm or even a heavy rain as the head of grain became heavier, leading to harvest losses. The new short, stiff-strawed varieties could support a much larger head of grain without toppling over.

In the 1930s, plant breeders in the United States were raising yields of corn with high-yielding hybrid varieties. It was discovered that, with the right combination of parent stock, hybridization could dramatically increase yields. As the new hybrids spread in the United States, corn yields began to climb, quintupling between 1940 and 2010.

In contrast to wheat and rice, where dwarfing held the key to raising yields, corn breeders have worked in recent decades to develop hybrids that would tolerate crowding, enabling farmers to grow more corn plants per acre. And since each plant typically produces one ear of corn, more plants mean more corn. A half-century ago farmers typically grew perhaps 10,000 corn plants per acre. Today states with adequate soil moisture have plant populations of 28,000 or more per acre.

Although people often ask about the potential to raise grain yields using genetic modification, success has thus far been limited. This is largely because plant breeders using traditional approaches were successful in doing almost everything plant scientists could think of to raise yields, leaving little potential for doing so through genetic modification.

The tripling of world irrigated area since 1950 has also helped raise grain yields by helping high-yielding crops realize their full genetic potential. And because irrigation removes moisture constraints, it also facilitates the greater use of fertilizer.

When German chemist Justus von Liebig demonstrated in 1847 that the major nutrients that plants removed from the soil could be applied in mineral form, he set the stage for the development of a new industry and a huge jump in world food production a century later. Of the 16 elements plants require to be properly nourished, three—nitrogen, phosphorus, and potassium—totally dominate the world fertilizer industry. World fertilizer use climbed from 14 million tons in 1950 to 177 million tons in 2010, helping to boost the world grain harvest nearly fourfold.

As the world economy evolved from being largely rural to being highly urbanized, the natural nutrient cycle was disrupted. In traditional rural societies, food is consumed locally, and human and animal waste is returned to the land, completing the nutrient cycle. But in highly urbanized societies, where food is consumed far from where it is produced, using fertilizer to replace the lost nutrients is the only practical way to maintain land productivity. It thus comes as no surprise that the growth in fertilizer use closely tracks the growth in urbanization, with much of it concentrated in the last 60 years.

The big three grain producers—China, India, and the United States—account for 58 percent of world fertilizer use. In the United States, the growth in fertilizer use came to an end in 1980, but—in an encouraging sign—grain yields have continued to climb. China's fertilizer use climbed rapidly in recent decades but has leveled off since 2007. While China uses nearly 50 million tons of fertilizer a year and India uses nearly 25 million tons, the United States uses only 20 million tons.

Given that China and the United States each produce roughly 400 million tons of grain, the grain produced per ton of fertilizer in the United States is more than double that of China. This is partly because American farmers are much more precise in matching application with need, but also partly because the United States is far and away the world's largest soybean producer. The soybean, being a legume, fixes nitrogen in the soil that can be used by subsequent crops. U.S. farmers regularly plant corn and soybeans in a two-year rotation, thus reducing the amount of nitrogen fertilizer that has to be applied for the corn.

In most countries outside of sub-Saharan Africa, grain yields have doubled, tripled, or even quadrupled. Aside from having some of the world's inherently least fertile soils and a largely semiarid climate, sub-Saharan Africa lacks the infrastructure and modern inputs needed to support modern agriculture.

Recent experience in Malawi, however, illustrates the potential for improvement. After a drought in 2005, many of the country's 13 million people were left hungry or starving. In response, the government issued coupons to small farmers, entitling them to 200 pounds of fertilizer at a greatly reduced price and free packets of improved seed corn, the national food staple. Funded partly by outside donors, this fertilizer and seed subsidy program helped nearly double Malawi's corn harvest within two years, enabling it to export grain and boost farmers' incomes. With economic incentives and access to modern inputs, principally higher-yielding seed and fertilizer, farmers in sub-Saharan Africa can easily double yields.

At 10 tons per hectare, U.S. corn yields are the highest of any major grain anywhere. In Iowa, with its deep soils and near-ideal climate for corn, some counties harvest up to 13 tons per hectare. In China, yields of each of its "big three" grains—wheat, rice, and corn—now range between 4 and 6 tons. Wheat yields in India have more than quadrupled since 1950, climbing to 3 tons per hectare. Remember, all grain yields in India are lower than in the United States, Europe, or China because India is close to the equator, where yields are restricted by short day length.

Rising yields are the key to expanding the grain harvest. Since 1950, over 93 percent of world grain harvest growth has come from raising yields. Expanding area accounts for the other 7 percent.

Impressive though the growth is over the last 60 years, the pace has slowed during the last two decades. Between 1950 and 1990, the world grain yield increased by an average of 2.2 percent a year. From 1990 to 2011, the annual rise slowed to 1.3 percent. In some agriculturally advanced countries, the dramatic climb in yields has come to an end as yields have plateaued.

For example, the rice yield per hectare in Japan, after climbing for more than a century, has not increased at all over the last 17 years. It is not that Japanese farmers do not want to continue raising their rice yields. They do. With a domestic support price far above the world market price, raising yields in Japan is highly profitable. The problem is that Japan's farmers are already using all the technologies available to raise land productivity.

Like Japan, South Korea's rice yield also has plateaued. Interestingly, it plateaued at almost exactly the same level as the rice yield in Japan did, and while Japan's plateauing began in 1994, South Korea's began in 1996. The constraints on rice yields appear to be essentially the same in both countries. Yields there have hit a glass ceiling, a limit that is apparently imposed by day length, solar intensity, and, ultimately, the constraints of photosynthetic efficiency. Japan and South Korea together produce 12 million tons of rice annually, 3 percent of the world rice harvest.

A similar situation is developing with wheat in Europe. In France, Germany, and the United Kingdom, wheat yields have been flat for more than a decade. Eight tons per hectare appears to be the biological upper limit for wheat yields in the United Kingdom and Germany. For

France, located several degrees southward, and thus with somewhat shorter summer days, it is closer to 7 tons. (See Figure 7–1.) These three countries—France, Germany, and the United Kingdom—together produce 80 million tons of wheat, roughly 12 percent of the world wheat harvest.

One thing that has become clear is that grain yield per hectare, like any biological growth process, cannot continue rising indefinitely. It has its limits. Once we remove nutrient constraints by applying fertilizer and we remove soil moisture constraints wherever possible by irrigating, then it is the potential of photosynthesis and the local climate that ultimately limits crop yields.

Thus far the countries where rice or wheat yields have plateaued are medium-sized ones. What happens when grain yields plateau in some of the larger countries? Among the crops that are of particular concern are rice and wheat in China, which has the world's largest harvest of both, and corn in the United States, by far its largest producer. In each of these situations, the current yield is

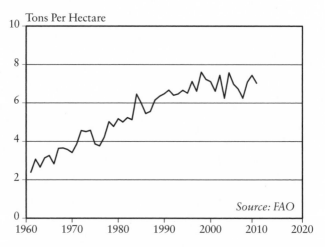

Figure 7–1. *Wheat Yields in France, 1961–2010*

quite high and may not continue rising much longer.

Rice yields in China are now very close to those in Japan. (See Figure 7–2.) Unless Chinese farmers can somehow surpass their Japanese counterparts, which seems unlikely, China's rice yields appear about to plateau. If China hits the glass ceiling for its rice yields, then one third of the world's rice would be produced in three countries (Japan, South Korea, and China) that can no longer raise land productivity or expand the area in rice. Future gains in the rice harvest would have to come from countries that account for the remaining two thirds of the world's rice harvest, but some of these could be approaching their own glass ceilings.

China's wheat may also be getting close to the glass ceiling. There are no longer many additional steps that China's farmers can take to raise yields. In a country that is already using twice as much fertilizer as the United States, it is highly unlikely that using more fertilizer will raise yields. There is little to no potential for expanding

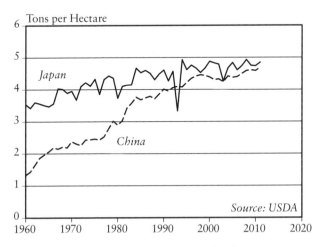

Figure 7–2. *Rice Yields in Japan and China, 1960–2011*

irrigation. Thus, the rapid rise in rice and wheat yields in recent decades in China may largely have run its course.

If China's wheat yields plateau, along with those of the three leading producers in Western Europe, nearly 30 percent of the world's wheat harvest would be grown in countries that may not be able to achieve any future meaningful gains in their output.

Corn yields in the United States, currently at 10 tons per hectare, have not yet plateaued. But although corn is more photosynthetically efficient than other grains, it too has its biological limits. If the United States is approaching the point where it can no longer systematically raise corn yields, it would very much affect the global corn prospect, since the United States accounts for 40 percent of the world harvest.

As yields continue to rise, more and more countries will edge ever closer to their glass ceilings. At the same time, the earth's rising temperature is making it more difficult to sustain a steady rise in grain yields. Unfortunately, these are not the only emerging constraints on efforts to expand food production.

Data, endnotes, and additional resources can be found at Earth Policy Institute, www.earth-policy.org.

8

Rising Temperatures,
Rising Food Prices

Agriculture as it exists today developed over 11,000 years
of rather remarkable climate stability. It has evolved to
maximize production within that climate system. Now,
suddenly, the climate is changing. With each passing year,
the agricultural system is becoming more out of sync with
the climate system.

In generations past, when there was an extreme weather
event, such as a monsoon failure in India, a severe drought
in Russia, or an intense heat wave in the U.S. Corn Belt, we
knew that things would shortly return to normal. But today there is no "normal" to return to. The earth's climate
is now in a constant state of flux, making it both unreli-
able and unpredictable.

Since 1970, the earth's average temperature has risen
more than 1 degree Fahrenheit. (See Figure 8–1.) If we
continue with business as usual, burning ever more oil,
coal, and natural gas, it is projected to rise by some 11
degrees Fahrenheit (6 degrees Celsius) by the end of this
century. The rise will be uneven. It will be much greater in
the higher latitudes than in the equatorial regions, greater over land than over oceans, and greater in continental
interiors than in coastal regions.

As the earth's temperature rises, it affects agriculture in

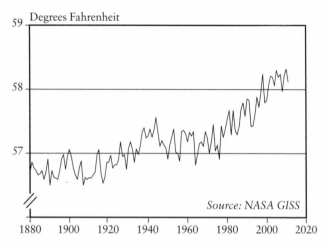

Figure 8–1. *Average Global Temperature 1880–2011*

many ways. High temperatures interfere with pollination and reduce photosynthesis of basic food crops. The most vulnerable part of a plant's life cycle is the pollination period. Of the world's three food staples—corn, wheat, and rice—corn is particularly vulnerable. In order for corn to reproduce, pollen must fall from the tassel to the strands of silk that emerge from the end of each ear. Each of these silk strands is attached to a kernel site on the cob. If the kernel is to develop, a grain of pollen must fall on the silk strand and then journey to the kernel site where fertilization takes place. When temperatures are uncommonly high, the silk strands quickly dry out and turn brown, unable to play their role in the fertilization process.

When it comes to rice, the effects of temperature on pollination have been studied in detail in the Philippines. Scientists there report that the pollination of rice falls from 100 percent at 93 degrees Fahrenheit (34 degrees Celsius) to near zero at 104 degrees, leading to crop failure.

High temperatures can also dehydrate plants. When a

corn plant curls its leaves to reduce exposure to the sun, photosynthesis is reduced. And when the stomata on the underside of the leaves close to reduce moisture loss, carbon dioxide (CO_2) intake is also reduced, further restricting photosynthesis. At elevated temperatures, the corn plant, which under ideal conditions is so extraordinarily productive, goes into thermal shock.

In a study of local ecosystem sustainability, Mohan Wali and his colleagues at Ohio State University noted that as temperature rises, photosynthetic activity in plants increases until the temperature reaches 68 degrees Fahrenheit. The rate of photosynthesis then plateaus until the temperature reaches 95 degrees Fahrenheit. Beyond this point it declines, until at 104 degrees Fahrenheit, photosynthesis ceases entirely.

All of these changes affect crop yields. Crop ecologists in several countries have been focusing on the precise relationship between temperature and crop yields. Their findings suggest a rule of thumb that a 1-degree-Celsius rise in temperature above the norm during the growing season lowers wheat, rice, and corn yields by 10 percent. Some of the most comprehensive research on this topic comes from the International Rice Research Institute in the Philippines. Crop yields from experimental field plots of irrigated rice dropped by 10 percent with a 1-degree-Celsius rise in temperature. The scientists concluded that "temperature increases due to global warming will make it increasingly difficult to feed Earth's growing population."

Stanford University scientists David Lobell and Gregory Asner conducted an empirical analysis of the effect of temperature on U.S. corn and soybean yields. They found that higher temperatures during the growing season had an even greater effect on yields of these crops than many scientists had reckoned. Using data for 1982–98 from 618 counties for corn and 444 counties for soybeans, they

concluded that for each 1-degree-Celsius rise in temperature, yields of each crop declined by 17 percent. This study
suggests that the earlier rule of thumb that a 1-degree-
Celsius rise in temperature would reduce yields by 10 percent could be conservative.

The earth's rising temperature also affects crop yields
indirectly via the melting of mountain glaciers. As the
larger glaciers shrink and the smaller ones disappear, the
ice melt that sustains rivers, and the irrigation systems
dependent on them, will diminish. In early 2012, a release
from the University of Zurich's World Glacier Monitoring
Service indicated that 2010 was the twenty-first consecutive year of glacier retreat. They also noted that glaciers
are now melting at least twice as fast as a decade ago.

Mountain glaciers are melting in the Andes, the Rocky
Mountains, the Alps, and elsewhere, but nowhere does
melting threaten world food security more than in the glaciers of the Himalayas and on the Tibetan Plateau that
feed the major rivers of India and China. It is the ice melt
that keeps these rivers flowing during the dry season. In the
Indus, Ganges, Yellow, and Yangtze River basins, where
irrigated agriculture depends heavily on rivers, the loss of
glacial-fed, dry-season flow will shrink harvests and could
create unmanageable food shortages.

In China, which is even more dependent than India on
river water for irrigation, the situation is particularly chal-
lenging. Chinese government data show that the glaciers
on the Tibetan Plateau that feed the Yellow and Yangtze Rivers are melting at a torrid pace. The Yellow River,
whose basin is home to 153 million people, could experience a large dry-season flow reduction. The Yangtze River,
by far the larger of the two, is threatened by the disappear-
ance of glaciers as well. The basin's 586 million people
rely heavily on rice from fields irrigated with its water.

Yao Tandong, one of China's leading glaciologists,

predicts that two thirds of China's glaciers could be gone by 2060. "The full-scale glacier shrinkage in the plateau region," Yao says, "will eventually lead to an ecological catastrophe."

The world has never faced such a predictably massive threat to food production as that posed by the melting mountain glaciers of Asia. China and India are the world's top two wheat producers, and they also totally dominate the rice harvest.

Agriculture in the Central Asian countries of Afghanistan, Kazakhstan, Kyrgyzstan, Tajikistan, Turkmenistan, and Uzbekistan depends heavily on snowmelt from the Hindu Kush, Pamir, and Tien Shan mountain ranges for irrigation water. Nearby Iran gets much of its water from the snowmelt in the 18,000-foot-high Alborz Mountains between Tehran and the Caspian Sea. The glaciers in these ranges also appear vulnerable to rising temperatures.

In the Andes, a number of small glaciers have already disappeared, such as the Chacaltaya in Bolivia and Cotacachi in Ecuador. Within a couple of decades, numerous other glaciers are expected to follow suit, disrupting local hydrological patterns and agriculture. For places that rely on glacial melt for household and irrigation use, this is not good news.

Peru, which stretches some 1,100 miles along the vast Andean mountain range and is the site of 70 percent of the earth's tropical glaciers, is in trouble. Its glaciers, which feed the many Peruvian rivers that supply water to the cities in the semiarid coastal regions, have lost 22 percent of their area. Ohio State University glaciologist Lonnie Thompson reported in 2007 that the Quelccaya Glacier in southern Peru, which was retreating by 6 meters per year in the 1960s, was by then retreating by 60 meters annually. In an interview with *Science News* in early 2009, he said, "It's now retreating up the mountainside by about

18 inches a day, which means you can almost sit there and watch it lose ground."

Many of Peru's farmers irrigate their wheat, rice, and potatoes with the river water from these disappearing glaciers. During the dry season, farmers are totally dependent on irrigation water. For Peru's 30 million people, shrinking glaciers could mean shrinking harvests.

Throughout the Andean region, climate change is contributing to water scarcity. Barbara Fraser writes in *The Daily Climate* that "experts predict that climate change will exacerbate water scarcity, increasing conflicts between competing users, pitting city dwellers against rural residents, people in dry lands against those in areas with abundant rainfall and Andean mining companies against neighboring farm communities."

In the southwestern United States, the Colorado River—the region's primary source of irrigation water—depends on snowfields in the Rockies for much of its flow. California, in addition to depending heavily on the Colorado, relies on snowmelt from the Sierra Nevada range in the eastern part of the state. Both the Sierra Nevada and the coastal range supply irrigation water to California's Central Valley, the country's fruit and vegetable basket.

With the continued heavy burning of fossil fuels, global climate models project a 70-percent reduction in the amount of snow pack for the western United States by mid-century. The Pacific Northwest National Laboratory of the U.S. Department of Energy did a detailed study of the Yakima River Valley, a vast fruit-growing region in Washington State. It projected progressively heavier harvest losses as the snow pack shrinks, reducing irrigation water flows.

Even as the melting of glaciers threatens dry-season river flows, the melting of mountain glaciers and of the Greenland and Antarctic ice sheets is raising sea level and thus threatening the rice-growing river deltas of Asia. If the Greenland ice sheet were to melt entirely, it would raise sea level 23 feet. The latest projections show sea level rising by up to 6 feet during this century. Such a rise would sharply reduce the rice harvest in Asia, home to over half the world's people. Even half that rise would inundate half the riceland in Bangladesh, a country of 152 million people, and would submerge a large part of the Mekong Delta, a region that produces half of Viet Nam's rice, leaving the many countries that import rice from it looking elsewhere.

In addition to the Gangetic and Mekong Deltas, numerous other rice-growing river deltas in Asia would be submerged in varying degrees by a 6-foot rise in sea level. It is not intuitively obvious that ice melting on a large island in the far North Atlantic could shrink the rice harvest in Asia, but it is true.

Scientists also expect higher temperatures to bring more drought—witness the dramatic increase in the land area affected by drought in recent decades. A team of scientists at the National Center for Atmospheric Research in the United States reported that the earth's land area experiencing very dry conditions expanded from well below 20 percent from the 1950s to the 1970s to closer to 25 percent in recent years. The scientists attributed most of the change to a rise in temperature and the remainder to reduced precipitation. The drying was concentrated in the Mediterranean region, East and South Asia, mid-latitude Canada, Africa, and eastern Australia.

A 2009 report published by the U.S. National Academy of Sciences reinforced these findings. It concluded that if atmospheric CO_2 climbs from the current level of 391 parts per million (ppm) to above 450 ppm, the world will face irreversible dry-season rainfall reductions in several regions. The study likened the conditions to those of the U.S. Dust Bowl era of the 1930s. Physicist Joe Romm, drawing on recent climate research, reports that "levels of aridity comparable to those in the Dust Bowl could stretch from Kansas to California by mid-century."

Rising temperatures also fuel wildfires. Anthony Westerling of Scripps Institution and colleagues found that the average wildfire season in the western United States has lengthened by 78 days from the period 1970–86 to 1987–2003 as temperatures increased an average 1.6 degrees Fahrenheit. Looking forward, researchers with the U.S. Department of Agriculture's Forest Service drew on 85 years of fire and temperature records to project that a 2.9-degree-Fahrenheit rise in summer temperature could double the area of wildfires in the 11 western states.

In addition to more widespread drought and more numerous wildfires, climate change brings more extreme heat waves. One of the most destructive of these came in the U.S. Midwest in 1988. Combined with drought, as most heat waves are, this one dropped the U.S. grain harvest from an annual average of 324 million tons in the preceding years to 204 million tons. Fortunately, the United States—the world's dominant grain supplier—had substantial stocks at that time that it could draw upon, allowing it to meet its export commitments. If such a drop were to occur today, when grain stocks are seriously depleted, there would be panic in the world grain market.

Another extreme heat wave came in Western Europe in the late summer of 2003. It claimed some 52,000 lives. France and Italy were hit hardest. And London experienced its first 100-degree-Fahrenheit temperature reading in its history. Fortunately the wheat crop was largely harvested when this late-summer heat wave began, so the losses in that sector were modest.

In the summer of 2010, Russia experienced an extraordinary heat wave unlike anything it had seen before. The July temperature in Moscow averaged a staggering 14 degrees Fahrenheit above the norm. High temperatures sparked wildfires, which caused an estimated $300 billion worth of damage to the country's forests. In addition to claiming nearly 56,000 lives, this heat wave reduced the Russian grain harvest from nearly 100 million tons to 60 million tons. Russia, which had been an exporting country, suddenly banned exports.

Close on the heels of these unprecedented high temperatures in Russia was the 2011 heat wave in Texas, a leading U.S. agricultural state. In Dallas, located in the Texas heartland, the average temperature reached 100 degrees Fahrenheit for 40 consecutive days, shattering all records. It also forced many farmers into bankruptcy. More than a million acres of crops were never harvested. Many ranchers in this leading cattle-producing state had to sell their herds. They had no forage, no water, and no choice. The heat and drought in Texas broke almost all records in the state's history for both intensity and duration. Agricultural damage was estimated to exceed $7 billion.

As the earth's temperature rises, scientists expect heat waves to be both more frequent and more intense. Stated otherwise, crop-shrinking heat waves will now become part of the agricultural landscape. Among other things, this means that the world should increase its carryover stocks of grain to provide adequate food security.

The continuing loss of mountain glaciers and the resulting reduced meltwater runoff could create unprecedented water shortages and political instability in some of the world's more densely populated countries. China, already struggling to contain food price inflation, could well see spreading social unrest if food supplies tighten.

For Americans, the melting of the glaciers on the Tibetan Plateau would appear to be China's problem. It is. But it is also a problem for the entire world. For low-income grain consumers, this melting poses a nightmare scenario.

If China enters the world market for massive quantities of grain, as it has already done for soybeans over the last decade, it will necessarily come to the United States—far and away the leading grain exporter. The prospect of 1.35 billion Chinese with rapidly rising incomes competing for the U.S. grain harvest, and thus driving up food prices for all, is not an attractive one.

In the 1970s, when tight world food supplies were generating unacceptable food price inflation in the United States, the government restricted grain exports. This is probably not an option today where China is concerned. Each month when the U.S. Treasury Department auctions off securities to cover the U.S. fiscal deficit, China is one of the big buyers. Now holding close to $1 trillion of U.S. debt, China has become the banker for the United States. Like it or not, Americans will be sharing their grain harvest with Chinese consumers. The idea that shrinking glaciers on the Tibetan Plateau could one day drive up food prices at U.S. supermarket checkout counters is yet another sign of how integrated our global civilization has become.

Data, endnotes, and additional resources can be found at Earth Policy Institute, www.earth-policy.org.

9

China and the Soybean Challenge

Some 3,000 years ago, farmers in eastern China domesticated the soybean. In 1765, the first soybeans arrived in North America, but they did not soon catch on as a crop. For 150 years or so the soybean languished as a curiosity in gardens.

Then in the late 1920s, a market for soybean oil began to develop, moving the soybean from the garden to the field. During the 1930s, soybean production in the United States climbed from 400,000 tons to over 2 million tons. And as growth in the demand for the oil gained momentum, soybean production jumped to over 8 million tons in 1950.

During the 1940s and early 1950s, the soybean crop was harvested and crushed primarily for the 20 percent of the bean that was oil. Then during the 1950s, the demand for meat, milk, and eggs climbed. With little new grassland to support expanding beef and dairy herds, farmers started feeding their animals more grain supplemented with soybean meal in order to produce more beef and milk. Farmers were already relying heavily on grain to produce pork, poultry, and eggs. By 1960 soybean meal had become the primary product of soybean crushing and oil the secondary one. For the first time, the value of the meal exceeded that of the oil, an early sign of things to come in the changing role of the soybean.

This rise in the demand for soybean meal reflected the discovery by animal nutritionists that combining 1 part soybean meal with 4 parts grain, usually corn, in feed rations would sharply boost the efficiency with which livestock and poultry converted grain into animal protein. This was the soybean's ticket to agricultural prominence, enabling it to join wheat, rice, and corn as one of the world's four leading crops. (See Figure 9–1.)

Although the soybean had originated in China, it found a welcome home in the United States. In its new role as a source of high-quality protein for mixing in animal feeds, it was destined to become an integral part of the U.S. farm economy.

After World War II, U.S. production of the soybean soared, bringing China's historical dominance of soybean production to an end. By 1960, the U.S. harvest was close

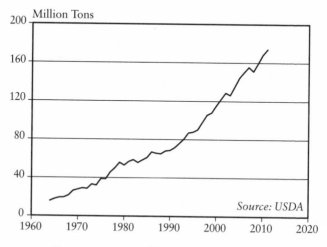

Figure 9–1. *World Soybean Meal Use for Feed, 1964–2011*

to triple that in China. By 1965, the United States was producing three fourths of the world's soybeans and accounting for virtually all the exports.

When world grain and soybean prices spiked in the mid-1970s following the 1972 Soviet crop failure, the United States—in an effort to curb domestic food price inflation—embargoed soybean exports. Japan, a leading importer, was soon looking for another supplier. And Brazil was looking for new crops to export. The rest is history, as Brazil became a leading soybean exporter.

Neighboring Argentina, a leading exporter of wheat and corn, also recognized the market potential for soybeans. Once the soybean gained a foothold in Argentina, production there expanded rapidly, making it the third of the big three soybean producers and exporters.

The main soybean producers today, in round numbers, are the United States at 80 million tons, Brazil at 70 million tons, and Argentina at 45 million tons. Together they account for over four fifths of world soybean production. China is a distant fourth at a mere 14 million tons. For six decades, the United States was both the leading producer and exporter of soybeans, but in 2011 Brazil's exports narrowly eclipsed those from the United States.

Throughout most of this period, the United States was also the leading soybean consumer. As recently as 1990, U.S. soybean consumption was quadruple that in China, but in 2008 China took the lead. By 2011 China was consuming 70 million tons of soybeans a year, well above the 50 million tons in the United States.

As China's appetite for meat, milk, and eggs has soared, so too has its use of soybean meal. And since nearly half the world's pigs are in China, the lion's share of soy use is in pig feed. Its fast-growing poultry industry is also dependent on soybean meal. In addition, China now uses large quantities of soy in feed for farmed fish.

Four numbers tell the story of the explosive growth of soybean consumption in China. In 1995, China was producing 14 million tons of soybeans and it was consuming 14 million tons. In 2011, it was still producing 14 million tons of soybeans—but it was consuming 70 million tons, meaning that 56 million tons had to be imported. (See Figure 9–2.)

China's neglect of soybean production reflects a political decision made in Beijing in 1995 to focus on being self-sufficient in grain. For the Chinese people, many of them survivors of the Great Famine of 1959–61, this was paramount. They did not want to be dependent on the outside world for their food staples. By strongly supporting grain production with generous subsidies and essentially ignoring soybean production, China increased its grain harvest rapidly while its soybean harvest languished.

Hypothetically, if China had chosen to produce all of the 70 million tons of soybeans it consumed in 2011,

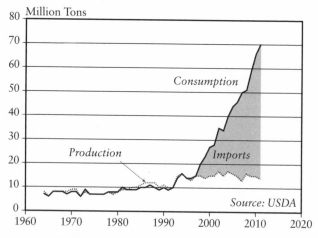

Figure 9–2. *Soybean Production, Consumption, and Imports in China, 1964–2011*

it would have had to shift one third of its grainland to soybeans, forcing it to import 160 million tons of grain—more than a third of its total grain consumption. Because of this failure to expand soybean production over the last 15 years or so, close to 60 percent of all soybeans entering international trade today go to China, making it far and away the world's largest importer. As more and more of China's 1.35 billion people move up the food chain, its soybean imports will almost certainly continue to climb.

Only one tenth of the soybeans used in China is consumed directly as food such as tofu and soy sauce. The other 90 percent is crushed, separating the oil and meal. In China, as elsewhere, the oil is a highly valued cooking oil and the meal is widely used in animal feed rations.

For the world as a whole the pattern of soybean consumption is similar. To most consumers, the soybean is an invisible food, one that is embodied in many of the products found in any refrigerator. Clearly, the soybean is far more pervasive in the human diet than the visual evidence would indicate.

The world demand for soybeans is increasing by some 7 million tons per year. It is being driven primarily by the 3 billion people who are moving up the food chain, consuming more grain- and soybean-intensive livestock products. Population growth is also driving up the demand for soybeans, either indirectly through the consumption of livestock products or directly through the consumption of tofu, miso, and tempeh. In the two leading consumers of soybeans, the United States and China, populations are growing by 3 million and 6 million per year, respectively. And finally, an increasing demand for soy oil for biodiesel is also ramping up soybean use.

The principal effect of soaring world soybean consumption has been a restructuring of agriculture in the western hemisphere. In the United States there is now more land in soybeans than in wheat. In Brazil, the area in soybeans exceeds that of all grains combined. Argentina's soybean area is now close to double that of all grains combined, putting the country dangerously close to becoming a soybean monoculture.

For the western hemisphere as a whole, the fast-expanding area planted to soybeans overtook that in wheat in 1994. As of 2010, there was more than twice as much land in soybeans as in wheat. The soybean eclipsed corn in area in 2001. (See Figure 9–3.)

Satisfying the climbing global demand for soybeans poses a huge challenge. Since the soybean is a legume, fixing atmospheric nitrogen in the soil, it is not as fertiliz-er-responsive as, say, corn, which has a ravenous appetite for nitrogen. And because the soy plant uses a portion of its metabolic energy to fix nitrogen, it has less energy to produce seed. This makes raising yields difficult.

Since the mid-twentieth century, the world grain harvest

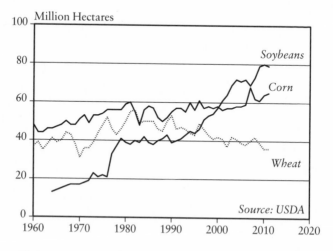

Figure 9–3. *Land in Corn, Wheat, and Soybeans in the Western Hemisphere, 1960–2011*

has nearly quadrupled, with most of this growth coming from the tripling of the grain yield per acre But the 16-fold increase in the global soybean harvest has come overwhelmingly from expanding the cultivated area. While the area expanded nearly sevenfold, the yield scarcely doubled. The world gets more soybeans primarily by planting more soybeans. Therein lies the problem.

The question then becomes, Where will the soybeans be planted? The United States is now using all of its available cropland and has no additional land that can be planted to soybeans. The only way to expand soybean acreage is by shifting land from other crops, such as corn or wheat.

In Brazil, new land for soybean production comes from the Amazon Basin or the *cerrado*, the savannah-like region to the south. Both the Amazon Basin and the *cerrado* are home to staggering levels of biodiversity, with many plant and animal species that can be found nowhere else on earth. Beyond this, both the regions store immense quantities of carbon, so new land clearing means not only lost biodiversity but also increased carbon emissions, exacerbating climate change for the entire world.

The Amazon Basin and the *cerrado* are also integral to the hydrological cycle. The Amazon rainforest recycles rainfall from the coastal regions to the continental interior, ensuring an adequate water supply for agriculture not only in Brazil's west and southwest but also in Paraguay and northern Argentina. And many of Brazil's rivers originate in the *cerrado*.

Unfortunately, land clearing has already taken a devastating toll on the Amazon Basin and the *cerrado*. Since 1970, the forested area in the Brazilian Amazon Basin has shrunk some 19 percent from its 400 million hectares. For the *cerrado*, it is estimated that roughly half of its original 200 million hectares has been lost. In both cases, soybean expansion has played a significant role.

In the *cerrado*, soybean farmers typically clear the land themselves. In the Amazon Basin, in contrast, they often purchase already deforested land from cattle ranchers. The ranchers in turn move further into the Amazon, clearing new land for their cattle. The cycle continues.

Some progress is being made in curbing land clearing in the Amazon Basin. Over the past decade in Mato Grosso, a large state on Brazil's agricultural frontier that produces nearly a third of the country's soybeans, deforestation slowed dramatically while soybean production increased rapidly.

Part of this reduction was due to government initiatives, such as restricting access to credit for deforesters, and a satellite monitoring system that provided information on when and where deforestation was occurring. This evidence in near real time proved to be a strong deterrent to deforestation. At the same time, a coalition of environmental groups pressured major soybean buyers to adopt a moratorium on purchasing soybeans produced on deforested land.

Unfortunately, if world soybean consumption continues to climb at a rapid rate, the economic pressures to clear more land could become intense. And if the additional land to meet the expanding demand is not in Brazil, where will it be? Where will the new land for soybeans come from?

Although the deforestation is occurring within Brazil, it is being driven by the worldwide growth in demand for meat, milk, and eggs. Put simply, saving the Amazon rainforest now depends on curbing the growth in demand for soybeans by stabilizing population worldwide as soon as possible. And for the world's more affluent population, it means eating less meat and thus slowing the growth in demand for soybeans. Against this backdrop, the recent downturn in U.S. meat consumption is welcome news.

10

The Global Land Rush

Between 2007 and mid-2008, world grain and soybean prices more than doubled. As food prices climbed everywhere, some exporting countries began to restrict grain shipments in an effort to limit food price inflation at home. Importing countries panicked. Some tried to negotiate long-term grain supply agreements with exporting countries, but in a seller's market, few were successful. Seemingly overnight, importing countries realized that one of their few options was to find land in other countries on which to produce food for themselves.

Looking for land abroad is not entirely new. Empires expanded through territorial acquisitions, colonial powers set up plantations, and agribusiness firms try to expand their reach. Agricultural analyst Derek Byerlee tracks market-driven investments in foreign land back to the mid-nineteenth century. During the last 150 years, large-scale agricultural investments from industrial countries concentrated primarily on tropical products such as sugarcane, tea, rubber, and bananas.

What is new now is the scramble to secure land abroad for more basic food and feed crops—including wheat, rice, corn, and soybeans—and for biofuels. These land acquisitions of the last several years, or "land grabs" as they are sometimes called, represent a new stage in the emerging geopolitics of food scarcity. They are occurring on a scale and at a pace not seen before.

Among the countries that are leading the charge to buy or lease land abroad, either directly through government entities or through domestically based agribusiness firms, are Saudi Arabia, South Korea, China, and India. Saudi Arabia's population has simply outrun its land and water resources. The country is fast losing its irrigation water and will soon be totally dependent on imports from the world market or overseas farming projects for its grain.

South Korea, which imports over 70 percent of its grain, is a major land investor in several countries. In an attempt to acquire 940,000 acres of farmland abroad by 2018 for corn, wheat, and soybean production, the Korean government will reportedly help domestic companies lease farmland or buy stakes in agribusiness firms in countries such as Cambodia, Indonesia, and Ukraine.

China, faced with aquifer depletion and the heavy loss of cropland to urbanization and industrial development, is also nervous about its future food supply. Although it was essentially self-sufficient in grain from 1995 onward, within the last few years China has become a leading grain importer. It is by far the top importer of soybeans, bringing in more than all other countries combined.

India, with a huge and growing population to feed, has also become a major player in land acquisitions. With irrigation wells starting to go dry, with the projected addition of 450 million people by mid-century, and with the prospect of growing climate instability, India too is worried about future food security.

Among the other countries jumping in to secure land abroad are Egypt, Libya, Bahrain, Qatar, and the United Arab Emirates (UAE). For example, in early 2012 Al Ghurair Foods, a company based in the UAE, announced it would lease 250,000 acres in Sudan for 99 years on which to grow wheat, other grains, and soybeans. The plan is that the resulting harvests will go to the UAE and other Gulf countries.

In tracking this worldwide land grab surge, accurate information has been difficult to find. Perhaps because of the politically sensitive nature of land grabs, separating rumor from reality remains a challenge. At the outset, the increasing frequency of news reports mentioning deals seemed to indicate that the phenomenon was growing, but no one was systematically aggregating and verifying data on this major agricultural development. Many groups have relied on GRAIN, a small nongovernmental organization (NGO) with a shoestring budget, and its compilations of media reports on land grabs. A much-anticipated World Bank report, first released in September 2010 and updated in January 2011, used GRAIN's online collection to aggregate land grab information, noting that GRAIN's was the only tracking effort that was global in scope.

In its report, the World Bank identified 464 land acquisitions that were in various stages of development between October 2008 and August 2009. It reported that production had begun on only one fifth of the announced projects, partly because many deals were made by land speculators. The report offered several other reasons for the slow start, including "unrealistic objectives, price changes, and inadequate infrastructure, technology, and institutions."

The amount of land involved was known for only 203 of the 464 projects, yet it still came to some 140 million acres—more than is planted in corn and wheat combined in the United States. Particularly noteworthy is that of the 405 projects for which commodity information was available, 21 percent were slated to produce biofuels and another 21 percent were for industrial or cash crops, such as rubber and timber. Only 37 percent of the projects involved food crops.

Nearly half of these land deals, and some two thirds of the land area, were in sub-Saharan Africa—partly because land is so cheap there compared with land in Asia. In a careful evidence-based analysis of land grabs in sub-Saharan Africa between 2005 and 2011, George Schoneveld from the Center for International Forestry Research reported that two thirds of the area acquired there was in just seven countries: Ethiopia, Ghana, Liberia, Madagascar, Mozambique, South Sudan, and Zambia. In Ethiopia, for example, an acre of land can be leased for less than $1 a year, whereas in land-scarce Asia it can easily cost $100 or more.

Nevertheless, the second-ranking region in land area involved was Southeast Asia, including Cambodia, Laos, the Philippines, and Indonesia. Countries have also sought land in Latin America, especially in Brazil and Argentina. The state-owned Chinese firm Chongqing Grain Group, for example, has reportedly begun harvesting soybeans on some 500,000 acres in Brazil's Bahia state for export to China. The company announced in early 2011 that as part of a multibillion-dollar investment package in Bahia, it would develop a soybean industrial park with facilities capable of crushing 1.5 million tons of soybeans a year.

Unfortunately, the countries selling or leasing their land for the production of agricultural commodities to be shipped abroad are typically poor and, more often than not, those where hunger is chronic, such as Ethiopia and South Sudan. Both of these countries are leading recipients of food from the U.N. World Food Programme. Some of these land acquisitions are outright purchases of land, but the overwhelming majority are long-term leases, typically 25 to 99 years.

In response to rising oil prices and a growing sense of oil insecurity, energy policies encouraging the production and use of biofuels are also driving land acquisitions. This results in either clearing new cropland or making existing cropland unavailable for food production. The European Union's renewable energy law requiring 10 percent of its transport energy to come from renewable sources by 2020, for instance, is encouraging agribusiness firms to invest in land to produce biofuels for the European market. In sub-Saharan Africa, many investors have planted jatropha (an oilseed-bearing shrub) and oil palm trees, both sources for biodiesel.

One company, U.K.-based GEM BioFuels, has leased 1.1 million acres in 18 communities in Madagascar on which to grow jatropha. At the end of 2010 it had planted 140,000 acres with this shrub. But by April 2012 it was reevaluating its Madagascar operations due to poor project performance. Numerous other firms planning to produce biodiesel from jatropha have not fared much better. The initial enthusiasm for jatropha is fading as yields are lower than projected and the economics just do not work out.

Sime Darby, a Malaysia-based company that is a big player in the world palm oil economy, has leased 540,000 acres in Liberia to develop oil palm and rubber plantations. It planted its first oil palm seedling on the acquired land in May 2011, and the company plans to have it all in production by 2030.

Thus we are witnessing an unprecedented scramble for land that crosses national boundaries. Driven by both food and energy insecurity, land acquisitions are now also seen as a lucrative investment opportunity. Fatou Mbaye of ActionAid in Senegal observes, "Land is quickly becoming the new gold and right now the rush is on."

Investment capital is coming from many sources, including investment banks, pension funds, university endowments, and wealthy individuals. Many large investment funds are incorporating farmland into their portfolios. In addition, there are now many funds dedicated exclusively to farm investments. These farmland funds generated a rate of return from 1991 to 2010 that was roughly double that from investing in gold or the S&P 500 stock index and seven times that from investing in housing. Most of the rise in farmland earnings has come since 2003.

Many investors are planning to use the land acquired, but there is also a large group of investors speculating in land who have neither the intention nor the capacity to produce crops. They sense that the recent rises in food prices will likely continue, making land even more valuable over the longer term. Indeed, land prices are on the rise almost everywhere.

Land acquisitions are also water acquisitions. Whether the land is irrigated or rainfed, a claim on the land represents a claim on the water resources in the host country. This means land acquisition agreements are a particularly sensitive issue in water-stressed countries.

In an article in *Water Alternatives*, Deborah Bossio and colleagues analyze the effect of land acquisition in Ethiopia on the demand for irrigation water and, in turn, its effect on the flow of the Nile River. Compiling data on 12 confirmed projects with a combined area of 343,000 acres, they calculate that if this land is all irrigated, as seems likely, the irrigated area in the region would increase sevenfold. This would reduce the average annual flow of the Blue Nile by approximately 4 percent.

Acquisitions in Ethiopia, where most of the Nile's headwaters begin, or in the Sudans, which also tap water from the Nile, mean that Egypt will get less water, thus shrinking its wheat harvest and pushing its already heavy dependence on imported wheat even higher.

Massive land acquisitions raise many questions. Since productive land is not often idle in the countries where the land is being acquired, the agreements mean that many local farmers and herders will simply be displaced. Their land may be confiscated or it may be bought from them at a price over which they have little say, leading to the public hostility that often arises in host countries.

In addition, the agreements are almost always negotiated in secret. Typically only a few high-ranking officials are involved, and the terms are often kept confidential. Not only are key stakeholders such as local farmers not at the negotiating table, they often do not even learn about the agreements until after the papers are signed and they are being evicted. Unfortunately, it is often the case in developing countries that the state, not the farmer, has formal ownership of the land. Against this backdrop, the poor can easily be forced off the land by the government.

The displaced villagers will be left without land or livelihoods in a situation where agriculture has become highly mechanized, employing few people. The principal social effect of these massive land acquisitions may well be an increase in the ranks of the world's hungry.

The Oakland Institute, a California-based think tank, reports that Ethiopia's huge land leases to foreign firms have led to "human rights violations and the forced relocation of over a million Ethiopians." Unfortunately, since the Ethiopian government is pressing ahead with its land lease program, many more villagers are likely to be forcibly displaced.

In a landmark article on African land grabs in the *Observer*, John Vidal quotes an Ethiopian, Nyikaw Ochalla, from the Gambella region: "The foreign companies are arriving in large numbers, depriving people of land they have used for centuries. There is no consultation with the indigenous population. The deals are done secretly. The only thing the local people see is people coming with lots of tractors to invade their lands." Referring to his own village, where an Indian corporation is taking over, Ochalla says, "Their land has been compulsorily taken and they have been given no compensation. People cannot believe what is happening."

Hostility of local people to land grabs is the rule, not the exception. China, for example, signed an agreement with the Philippine government in 2007 to lease 2.5 million acres of land on which to produce crops that would be shipped home. Once word leaked out, the public outcry—much of it from Filipino farmers—forced the government to suspend the agreement. A similar situation developed in Madagascar, where a South Korean firm, Daewoo Logistics, had pursued rights to more than 3 million acres of land, an area half the size of Belgium. This helped stoke a political furor that led to a change in government and cancellation of the agreement.

How productive will the land be that actually ends up being farmed? Given the level of agricultural skills and technologies likely to be used, in most cases robust gains in yields could be expected. As demonstrated in Malawi (see Chapter 7), simply applying fertilizer to nutrient-depleted soils where rainfall is adequate and using improved seed can easily double grain yields.

Perhaps the more important question is, What will be the effects on the local people? The Malawi program's approach of directly helping local farmers can dramatically expand food production, raise the income of villagers, reduce hunger, and earn foreign exchange—a win-win-win-win situation. This contrasts sharply with the lose-lose-lose situation accompanying land grabs—villagers lose their land, their food supply, and their livelihoods.

There will be some spectacular production gains in some countries; there will undoubtedly also be failures. Some projects have already been abandoned. Many more will be abandoned simply because the economics do not pan out. Long-distance farming, with the transportation and travel involved, can be costly, particularly when oil prices are high.

Overall, while announcements of new land acquisitions have been popping up with alarming frequency, the actual development of acquired land has been slow. Investors tend to focus on the costs of producing the crops without sufficiently considering the cost of building the modern agricultural infrastructure needed to support successful development of the tracts of acquired land. In most sub-Saharan African countries, there is little of this infrastructure, which means the cost to an investor of developing it can be overwhelming.

In some countries, it will take years to build the roads needed to both bring in agricultural inputs, such as fertilizer, and move the farm products out. Beyond this, there is a need for a local supply of either electric power or diesel fuel to operate irrigation pumps. A full-fledged farm equipment maintenance support system is needed, lest equipment is left idle while waiting for repair people and parts to come from afar. Maintaining a fleet of tractors, for example, requires not only trained mechanics but also an onsite inventory of things like tires and batteries. Grain elevators and grain dryers are essential for storing grain. Fertilizer and fuel storage facilities have to be constructed.

Another complicating factor is navigating the various governmental regulations and procedures. For example, as almost all the equipment and inputs needed in a modern farming operation have to be imported, this requires a familiarity with customs procedures. In addition, various permits may be required for such things as drilling irrigation wells, building irrigation canals, or tapping into the local electrical grid if one exists.

When Saudi Arabia decided to invest in cropland, it created King Abdullah's Initiative for Saudi Agricultural Investment Abroad, a program to facilitate land acquisitions and farming in other countries, including Sudan,

Egypt, Ethiopia, Turkey, Ukraine, Kazakhstan, the Philippines, Viet Nam, and Brazil. The Saudi Ministry of Commerce and Industry recently launched an inquiry to find out why things were moving at such a glacial pace. What they learned is that simply acquiring tracts of land abroad is only the first step. Modern agriculture depends on heavy investment in a supporting infrastructure, something that is costly even for the oil-rich Saudis.

There is also a huge knowledge deficit associated with launching new farming projects in countries where soils, climate, rainfall, insect pests, and crop diseases are far different from those in the investor country. There almost certainly will be unforeseen outbreaks of plant disease and insect infestations as new crops are introduced, particularly since so many of the land deals are in tropical and subtropical regions.

A lack of familiarity with the local environment brings with it a wide range of risks. The Indian firm Karuturi Global is the world's largest producer of cut roses, which it grows in Ethiopia, Kenya, and India for high-income markets. The company has recently entered the land rush, jumping at an offer in 2008 to farm up to 740,000 acres of land in Ethiopia's Gambella region. In 2011, the company planted its first corn crop in fertile land along the Baro River. Recognizing the possibility of flooding, Karuturi invested heavily in building dikes along the river. Unfortunately the dikes were not sufficient: 50,000 tons of corn were lost to flash flooding. Fortunately for Karuturi, the company was large enough to survive this heavy loss.

The bottom line is that investors face steep cost curves in bringing this land into production. Even though the land itself may be relatively inexpensive, the food grown under these conditions and shipped to home countries will be some of the most costly food ever produced.

Although the flurry of large-scale land acquisitions

began in 2008, as of 2012 there were only a few relatively small harvests to point to. The Saudis harvested their first rice crop in Ethiopia, albeit a very small one, in late 2008.

In 2009, South Korea's Hyundai Heavy Industries harvested some 4,500 tons of soybeans and 2,000 tons of corn on a 25,000-acre farm it took over from Russian owners, roughly 100 miles north of Vladivostok. Hyundai had planned to expand production rapidly to 100,000 tons of corn and soybeans by 2015. But in 2012 it anticipated producing only 9,000 tons of crops, putting it far behind schedule for reaching its 2015 goal. The advantage for Hyundai was that this was already a functioning farm. The supporting infrastructure was already in place. Yet even if Hyundai reaches its 100,000-ton goal, this will cover just 1 percent of South Korea's consumption of these commodities.

Another of the acquisitions that appears to be progressing is in South Sudan, where the Egyptian private equity company Citadel Capital has leased 260,000 acres for agriculture. In 2011 it began production with a 1,500-acre trial of chickpeas. The plan is to scale the area in chickpeas up to 130,000 acres in five years. The overall goal is to grow crops, eventually including corn and sorghum as well, for which there is a large local market and to produce them at well under the price of imports. This particular project is apparently intended to produce for local consumption. Unfortunately, this is not the case for the great majority of foreign acquisitions.

Land acquisitions, whether to produce food, biofuels, or other crops, raise questions about who will benefit. When virtually all the inputs—the farm equipment, the fertilizer, the pesticides, the seeds—are brought in from abroad and all the output is shipped out of the country, this contributes little to the local economy and nothing to the local food supply. These land grabs are not only benefiting the rich, they are doing so at the expense of the poor.

One of the most difficult variables to evaluate is political stability in the countries where land acquisitions are occurring. If opposition political parties come into office, they may cancel the agreements, arguing that they were secretly negotiated without public participation or support. Land acquisitions in South Sudan and the Democratic Republic of the Congo, both among the top failing states, are particularly risky. Few things are more likely to fuel insurgencies than taking land away from people. Agricultural equipment is easily sabotaged. If ripe fields of grain are torched, they burn quickly.

In Ethiopia, local opposition to land grabs appears to be escalating from protest to violence. In late April 2012, gunmen in the Gambella region attacked workers on land acquired by Saudi billionaire Mohammed al-Amoudi for rice production. They reportedly killed five workers and wounded nine others. Al-Amoudi's firm Saudi Star Agricultural Development was growing rice on just 860 acres of its 24,700-acre lease as of mid-2012, but it intends eventually to obtain another 716,000 acres in the region, with much of the rice harvest to be exported to Saudi Arabia.

The World Bank, working with the U.N. Food and Agriculture Organization and other related agencies, has formulated a set of principles governing land acquisitions. These guiding principles are well conceived, but unfortunately there is no mechanism to enforce them. The Bank does not seem willing to challenge the basic argument of those acquiring land, who continue to insist that it will benefit the people who live in the host countries.

Land acquisitions are being fundamentally challenged by a coalition of more than 100 NGOs, some national and others international. These groups argue that the world does not need big corporations bringing large-scale, heavily mechanized, capital-intensive agriculture into developing countries. Instead, these countries need international support for local village-level farming centered on labor-intensive family farms that produce for local and regional markets and that create desperately needed jobs.

As land and water become scarce, as the earth's temperature rises, and as world food security deteriorates, a dangerous geopolitics of food scarcity is emerging. The conditions giving rise to this have been in the making for several decades, but the situation has come into sharp focus only in the last few years. The land acquisitions discussed here are an integral part of a global power struggle for control of the earth's land and water resources.

Data, endnotes, and additional resources can be found at Earth Policy Institute, www.earth-policy.org.

11

Can We Prevent a Food Breakdown?

World agriculture is now facing challenges unlike any before. Producing enough grain to make it to the next harvest has challenged farmers ever since agriculture began, but now the challenge is deepening as new trends—falling water tables, plateauing grain yields, and rising temperatures—join soil erosion to make it difficult to expand production fast enough. As a result, world grain carryover stocks have dropped from an average of 107 days of consumption a decade or so ago to 74 days in recent years.

World food prices have more than doubled over the last decade. Those who live in the United States, where 9 percent of income goes for food, are largely insulated from these price shifts. But how do those who live on the lower rungs of the global economic ladder cope? They were already spending 50–70 percent of their income on food. Many were down to one meal a day before the price rises. Now millions of families routinely schedule one or more days each week when they will not eat at all.

What happens with the next price surge? Belt tightening has worked for some of the poorest people so far, but this cannot go much further. Spreading food unrest will likely lead to political instability. We could see a breakdown of political systems. Some governments may fall.

As food supplies have tightened, a new geopolitics of food has emerged—a world in which the global competition for land and water is intensifying and each country is fending for itself. We cannot claim that we are unaware of the trends that are undermining our food supply and thus our civilization. We know what we need to do.

There was a time when if we got into trouble on the food front, ministries of agriculture would offer farmers more financial incentives, like higher price supports, and things would soon return to normal. But responding to the tightening of food supplies today is a far more complex undertaking. It involves the ministries of energy, water resources, transportation, and health and family planning, among others. Because of the looming specter of climate change that is threatening to disrupt agriculture, we may find that energy policies will have an even greater effect on future food security than agricultural policies do. In short, avoiding a breakdown in the food system requires the mobilization of our entire society.

On the demand side of the food equation, there are four pressing needs—to stabilize world population, eradicate poverty, reduce excessive meat consumption, and reverse biofuels policies that encourage the use of food, land, or water that could otherwise be used to feed people. We need to press forward on all four fronts at the same time.

The first two goals are closely related. Indeed, stabilizing population depends on eliminating poverty. Even a cursory look at population growth rates shows that the countries where population size has stabilized are virtually all high-income countries. On the other side of the coin, nearly all countries with high population growth rates are on the low end of the global economic ladder.

The world needs to focus on filling the gap in reproductive health care and family planning while working to eradicate poverty. Progress on one will reinforce progress on the other. Two cornerstones of eradicating poverty are making sure that all children—both boys and girls—get at least an elementary school education and rudimentary health care. And the poorest countries need a school lunch program, one that will encourage families to send children to school and that will enable them to learn once they get there.

Shifting to smaller families has many benefits. For one, there will be fewer people at the dinner table. It comes as no surprise that a disproportionate share of malnutrition is found in larger families.

At the other end of the food spectrum, a large segment of the world's people are consuming animal products at a level that is unhealthy and contributing to obesity and cardiovascular disease. The good news is that when the affluent consume less meat, milk, and eggs, it improves their health. When meat consumption falls in the United States, as it recently has, this frees up grain for direct consumption. Moving down the food chain also lessens pressure on the earth's land and water resources. In short, it is a win-win-win situation.

Another initiative, one that can quickly lower food prices, is the cancellation of biofuel mandates. There is no social justification for the massive conversion of food into fuel for cars. With plug-in hybrids and all-electric cars coming to market that can run on local wind-generated electricity at a gasoline-equivalent cost of 80¢ per gallon, why keep burning costly fuel at four times the price?

On the supply side of the food equation, we face several challenges, including stabilizing climate, raising water productivity, and conserving soil. Stabilizing climate is not easy, but it can be done if we act quickly. It will take a huge cut in carbon emissions, some 80 percent within a decade, to give us a chance of avoiding the worst consequences of climate change. This means a wholesale restructuring of the world energy economy.

The easiest way to do this is to restructure the tax system. The market has many strengths, but it also has some dangerous weaknesses. It readily captures the direct costs of mining coal and delivering it to power plants. But the market does not incorporate the indirect costs of fossil fuels in prices, such as the costs to society of global warming. Sir Nicholas Stern, former chief economist at the World Bank, noted when releasing his landmark study on the costs of climate change that climate change was the product of a massive market failure.

The goal of restructuring taxes is to lower income taxes and raise carbon taxes so that the cost of climate change and other indirect costs of fossil fuel use are incorporated in market prices. If we can get the market to tell the truth, the transition from coal and oil to wind, solar, and geothermal energy will move very fast. If we remove the massive subsidies to the fossil fuel industry, we will move even faster.

Although to some people this energy transition may seem farfetched, it is moving ahead, and at an exciting pace in some countries. For example, four states in northern Germany now get at least 46 percent of their electricity from wind. For Denmark, the figure is 26 percent. In the United States, both Iowa and South Dakota now get one fifth of their electricity from wind farms. Solar power in Europe can now satisfy the electricity needs of some 15 million households. Kenya now gets one fifth of its electricity from geothermal energy. And Indonesia is shooting for 9,500 megawatts of geothermal generating capacity by 2025, which would meet 56 percent of current electricity needs.

In addition to the carbon tax, we need to reduce dependence on the automobile by upgrading public transportation worldwide to European standards. Where cars are used, the emphasis should be on electrifying them. The world has already partly electrified its passenger rail systems. As we shift from traditional oil-powered engines to plug-in hybrids and all-electric cars, we can substitute electricity from renewable sources for oil. In the meantime, as the U.S. automobile fleet, which peaked in 2008, shrinks, U.S. gasoline use will continue the decline of recent years. This decline, in the country that consumes more gasoline than the next 16 countries combined, is a welcome new trend.

Along with stabilizing climate, another key component to avoiding a breakdown in the food system is to raise water productivity. This could be patterned after the worldwide effort launched over a half-century ago to raise cropland productivity. This extraordinarily successful earlier endeavor tripled the world grain yield per acre between 1950 and 2011.

Raising water productivity begins with agriculture, simply because 70 percent of all water use goes to irrigation. Some irrigation technologies are much more efficient than others. The least efficient are flood and furrow irrigation. Sprinkler irrigation, using the center-pivot systems that are widely seen in the crop circles in the western U.S. Great Plains, and drip irrigation are far more efficient. The advantage of drip irrigation is that it applies water very slowly at a rate that the plants can use, losing little to evaporation. It simultaneously raises yields and reduces water use. Because it is labor-intensive, it is used primarily to produce high-value vegetable crops or in orchards.

Another option is to encourage the use of more water-efficient crops, such as wheat, instead of rice. Egypt, for example, limits the production of rice. China banned rice production in the Beijing region. Moving down the food chain also saves water.

Although urban water use is relatively small compared with that used for irrigation, cities too can save water. Some cities now are beginning to recycle much if not most of the water they use. Singapore, whose freshwater supplies are severely restricted by geography, relies on a graduated water tax—the more water you use, the more you pay per gallon—and an extensive water recycling program to meet the needs of its 5 million residents.

The key to raising water use efficiency is price policy. Because water is routinely underpriced, especially that used for irrigation, it is used wastefully. Pricing water to encourage conservation could lead to huge gains in water use efficiency, in effect expanding the supply that could in turn be used to expand the irrigated area.

The third big supply-side challenge after stabilizing climate and raising water productivity is controlling soil erosion. With topsoil blowing away at a record rate and two huge dust bowls forming in Asia and Africa, stabilizing soils will take a heavy investment in conservation measures. Perhaps the best example of a large-scale effort to reduce soil erosion came in the 1930s, after a combination of overplowing and land mismanagement created a dust bowl that threatened to turn the U.S. Great Plains into a vast desert.

In response to this traumatic experience, the United States introduced revolutionary changes in agricultural practices, including returning highly erodible land to grass, terracing, planting tree shelterbelts, and strip cropping (planting wheat on alternative strips with fallowed land each year). The government also created a remarkably successful new agency in the U.S. Department of Agriculture—the Soil Conservation Service—whose sole responsibility was to manage and protect soils in the United States.

Another valuable tool in the soil conservation tool kit is no-till farming. Instead of the traditional practice of plowing land and discing or harrowing it to prepare the seedbed, and then using a mechanical cultivator to control weeds in row crops, farmers simply drill seeds directly through crop residues into undisturbed soil, controlling weeds with herbicides when necessary. In addition to reducing erosion, this practice retains water, raises soil organic matter content, and greatly reduces energy use for tillage.

In the United States, the no-till area went from 7 million hectares in 1990 to 26 million hectares (67 million acres) in 2007. Now widely used in the production of corn and soybeans, no-till agriculture has spread rapidly in the western hemisphere, covering 26 million hectares each in Brazil and Argentina and 13 million hectares in Canada. Australia, with 17 million hectares, rounds out the five leading no-till countries.

If we pursue the initiatives on both sides of the food equation as just outlined, we can rebuild world grain stocks to the level needed to improve food security. Since we no longer have idled cropland to bring back into production, our only cushion in the event of a disastrous world harvest is these carryover stocks.

No one knows for sure what level of stocks would be adequate today, but if stocks equal to 70 days of grain consumption were sufficient 40 years ago, then today we should plan on stocks equal to at least 110 days of consumption to take into account the more extreme weather events that come with climate change.

These initiatives do not constitute a menu from which to pick and choose. We need to take all these actions simultaneously. They reinforce each other. We will not likely be able to stabilize population unless we eradicate poverty. We will not likely be able to restore the earth's natural systems without stabilizing population and stabilizing climate. Nor can we eradicate poverty without reversing the decline of the earth's natural systems.

Achieving all these goals to reduce demand and increase supply requires that we redefine security. We have inherited a definition of security from the last century, a century dominated by two world wars and a cold war, that is almost exclusively military in focus. When the term national security comes up in Washington, people automatically think of expanded military budgets and more-advanced weapon systems. But armed aggression is no longer the principal threat to our future. The overriding threats in this century are climate change, population growth, spreading water shortages, rising food prices, and politically failing states.

It is no longer possible to separate food security and security more broadly defined. It is time to redefine security not just in an intellectual sense but also in a fiscal sense. We have the resources we need to fill the family planning gap, to eradicate poverty, and to raise water productivity, but these measures require a reallocation of our fiscal resources to respond to the new security threats.

Beyond this, diverting a big chunk of the largely obsolete military budget into incentives to invest in rooftop solar panels, wind farms, geothermal power plants, and more energy-efficient lighting and household appliances would accelerate the energy transition. The incentives needed to jump-start this massive energy restructuring are large, but not beyond our reach. We can justify this expense simply by considering the potentially unbearable costs of continuing with business as usual.

We have to mobilize quickly. Time is our scarcest resource. Success depends on moving at wartime speed. It means, for example, transforming the world energy economy at a pace reminiscent of the restructuring of the U.S. industrial economy in 1942 following the Japanese surprise attack on Pearl Harbor on December 7, 1941.

On January 6, 1942, a month after the attack,

Franklin D. Roosevelt outlined arms production goals in his State of the Union address to the U.S. Congress and the American people. He said the United States was going to produce 45,000 tanks, 60,000 planes, and thousands of ships. Given that the country was still in a depression-mode economy, people wondered how this could be done. It required a fundamental reordering of priorities and some bold moves. The key to the 1942 industrial restructuring was the government's ban on the sale of cars, a ban that forced the auto industry into arms manufacturing. The ban lasted from April 1942 until the end of 1944. Every one of President Roosevelt's arms production goals was exceeded.

If the United States could totally transform its industrial economy in a matter of months in 1942, then certainly it can lead the world in restructuring the energy economy, stabilizing population, and rebuilding world grain stocks. The stakes now are even higher than they were in 1942. The challenge then was to save the democratic way of life, which was threatened by the fast-expanding empires of Nazi Germany and Imperial Japan. Today the challenge is to save civilization itself.

Scientists and many other concerned individuals have long sensed that the world economy had moved onto an environmentally unsustainable path. This has been evident to anyone who tracks trends such as deforestation, soil erosion, aquifer depletion, collapsing fisheries, and the increase in carbon dioxide in the atmosphere. What was not so clear was exactly where this unsustainable path would lead. It now seems that the most imminent effect will be tightening supplies of food. Food is the weak link in our modern civilization—just as it was for the Sumerians, Mayans, and many other civilizations that have come and gone. They could not separate their fate from that of their food supply. Nor can we.

The challenge now is to move our early twenty-first-century civilization onto a sustainable path. Every one of us needs to be involved. This is not just a matter of adjusting lifestyles by changing light bulbs or recycling newspapers, important though those actions are. Environmentalists have talked for decades about saving the planet, but now the challenge is to save civilization itself. This is about restructuring the world energy economy and doing it before climate change spirals out of control and before food shortages overwhelm our political system. And this means becoming politically active, working to reach the goals outlined above.

We all need to select an issue and go to work on it. Find some friends who share your concern and get to work. The overriding priority is redefining security and reallocating fiscal resources accordingly. If your major concern is population growth, join one of the internationally oriented groups and lobby to fill the family planning gap. If your overriding concern is climate change, join the effort to close coal-fired power plants. We can prevent a breakdown of the food system, but it will require a huge political effort undertaken on many fronts and with a fierce sense of urgency.

We all have a stake in the future of civilization. Many of us have children. Some of us have grandchildren. We know what we have to do. It is up to you and me to do it. Saving civilization is not a spectator sport.

Data, endnotes, and additional resources can be found at Earth Policy Institute, www.earth-policy.org.

Index

Acknowledgments

As I have noted before, if it takes a village to raise a child, then it takes the entire world to produce a book of this scope. We draw on the work of thousands of scientists and research teams throughout the world. The book process ends with another international effort, this one by the teams who will translate it into dozens of languages.

The Earth Policy research team is led by Janet Larsen, our Director of Research. Janet is also my alter ego, my best critic, and a sounding board for new ideas. In researching for this book, the team combed through thousands of research reports, articles, and books—gathering, organizing, and analyzing information.

J. Matthew Roney—now a veteran of several book cycles—Sara Rasmussen, Emily Adams, and Brigid Fitzgerald Reading anchored the research effort, feeding me a steady stream of valuable data, sometimes finding data that I did not know existed. Intern Hayley Moller provided strong support with both data gathering and fact checking. Every research team member also reviewed and critiqued the manuscript as it evolved. I am deeply grateful to each of them for their unflagging enthusiasm and dedication.

Reah Janise Kauffman, our Vice President, not only

manages the Institute, thus enabling me to concentrate on research, she also directs the Institute's outreach effort. This includes coordinating our worldwide network of publishers, organizing book tours, and working with the media. Her value to me is evidenced in our 26 years of working together as a team.

Millicent Johnson, our Manager of Publications Sales, handles our publications department and serves as our office quartermaster and librarian. Millicent cheerfully handles the thousands of book orders and takes pride in her one-day turnaround policy.

Julianne Simpson, Web Communications Coordinator, replaced Kristina Taylor, who moved on to new challenges while the book was in process. Julianne assisted Reah Janise in planning our web outreach strategy.

My thanks also to individuals who were particularly helpful in providing specific information: Ruba Al-Zu'bi, Aiguo Dai, Klaus Deininger, Rolf Derpsch, Bernard Francou, Harald Frederiksen, Claudia Ringler, William Ryerson, Laura Schafer, John Sheehy, Lakshmi Siva, Lonnie Thompson, Wang Tao, Hodan Farah Wells, and Yao Tandong.

For this book, we had only one outside reviewer, Maureen Kuwano Hinkle, who drew on her 26 years of work experience on environmental issues in agriculture when reviewing an early draft.

As always, we are in debt to our editor, Linda Starke, who brings over 30 years of international experience in editing environmental books and reports to the table. She has brought her sure hand to the editing of not only this book but all my books during this period.

The book was produced in record time thanks to the conscientious efforts of Elizabeth Doherty, who prepared the page proofs under a very tight deadline. And the index was quickly and ably prepared by Kate Mertes.

We are supported by a network of dedicated translators and publishers in some 30 languages, including all the major ones. In addition to English, our books appear in Arabic, Bulgarian, Catalan, Chinese, Croatian, Danish, Dutch, Farsi, French, German, Greek, Hindi, Hungarian, Indonesian, Italian, Japanese, Korean, Marathi, Norwegian, Polish, Portuguese, Romanian, Russian, Slovenian, Spanish, Swedish, Thai, Turkish, and Vietnamese.

These translations are the work of numerous environmentally committed individuals around the world. In Iran, the husband and wife team of Hamid Taravati and Farzaneh Bahar, both medical doctors, head an environmental NGO and translate EPI's publications into Farsi.

In China, Lin Zixin, who has arranged for the publication of my books in Chinese for nearly 30 years, enlisted the Shanghai Scientific & Technological Education Publishing House for *Plan B 4.0* and *World on the Edge*.

In Japan, we have the good fortune of having two organizations publish our works. Soki Oda, who started Worldwatch Japan some 20 years ago, has led our publication efforts, arranging excellent book promotional media interviews and public events. Our longtime supporter Junko Edahiro translated *World on the Edge*, which was published by Diamond.

Gianfranco Bologna, head of WWF-Italy, with whom I've had a delightful relationship for 34 years, arranges for publication of our books in Italy, mainly through Edizioni Ambiente, Italy's premier environmental publishing house.

In Romania, we are aided by former President Ion Iliescu, who started publishing our books 24 years ago when he headed the publishing house Editura Tehnica. Now he often personally launches our books in Romania, ably aided by Roman Chirila of Editura Tehnica.

In Turkey, TEMA—the leading environmental NGO,

which coordinates a national tree planting program—has published my books for many years.

In South Korea, Yul Choi, founder of the Korean Federation for Environmental Movement and now head of the Korea Green Foundation, has published my books and oversees their launching through Doyosae Books Co.

Most remarkable are the individuals who step forward out of seemingly nowhere to publish and promote our books. For instance, Lars and Doris Almström have now translated and published three editions in the *Plan B* series in Sweden. David Biro, a school teacher, has translated the same editions into Hungarian. Boksmia has produced two Norwegian editions, while Paper Tiger has produced three Bulgarian editions.

Pierre-Yves Longaretti and Philippe Vieille in France translated *Plan B 2.0* and then engaged publishing giant Calman-Lévy. For *World on the Edge*, they worked with publisher Rue de l'Echiquier.

In Brazil, Edoardo Rivetti teamed up with Ricardo Voltolini and Bradesco Bank to publish *Plan B 4.0* in record time. He has since also published *World on the Edge*.

The Spanish editions of the *Plan B* series have been spearheaded by Gilberto Rincon of the Centre of Studies for Sustainable Development in Colombia.

New publishers include Poduzetniśtvo Jakić doing a Croatian edition of *Plan B 4.0* and, for *World on the Edge*, Maurits Groen MGMC producing the Dutch edition, Chroniko producing the Greek edition, and Hanh Lien doing a Vietnamese translation.

We are also indebted to our funders, including the Foundation for the Carolinas; the United Nations Population Fund; and the Farview, Laney Thornton, Shenandoah, Wallace Genetic, and Weeden foundations. A very special thanks to retired Air Force Colonel Henry Ingwersen, who at age 90 donated his life savings to the Institute.

Earth Policy is also supported by individual donors. I would like in particular to thank the following for their major contributions: Patricia Anderson in memory of Ray Anderson, Charles Babbs, James Dehlsen, Junko Edahiro & Rich Oda, Sarah Epstein, Judith Gradwohl, Richard Haylock, Maureen Kuwano Hinkle, Sudhanshu Jain, Betty Wold Johnson, Giuseppe LaManna, William Mansfield, John McBride, Scott & Hella McVay, Mary Morse & James McBride, Sharon Nolting, Christopher Quirk, Michael Richtsmeier & Carol Daly, John B. Robbins, Roger & Vicki Sant, Peter Seidel, Sarah Sponheim, Emily Troemel, and Jeremy Waletzky.

Finally, my thanks to the team at W. W. Norton & Company: Amy Cherry, our book manager; Louise Mattarelliano, who put the book on a fast-track production schedule; Chin-Yee Lai, book jacket designer; Bill Rusin, Marketing Director; and Drake McFeely, President, with special thanks for his support. It is a delight to work with such a talented team and to have been published by W. W. Norton for more than 30 years.

And thanks to you, our readers. In the end, the success of this book depends on you and your active involvement in reaching its goals.

Lester R. Brown

About the Author

Lester R. Brown is President of Earth Policy Institute, a nonprofit, interdisciplinary research organization based in Washington, D.C., which he founded in May 2001. The purpose of the Earth Policy Institute is to provide a plan for sustaining civilization and a roadmap of how to get from here to there.

Brown has been described as "one of the world's most influential thinkers" by the *Washington Post*. The *Telegraph of Calcutta* called him "the guru of the environmental movement." In 1986, the Library of Congress requested his papers for its archives.

Some 30 years ago, Brown helped pioneer the concept of environmentally sustainable development, a concept embodied in Plan B. He was the Founder and President of the Worldwatch Institute during its first 26 years. During a career that started with tomato farming, Brown has authored or coauthored many books and been awarded 25 honorary degrees. With books in more than 40 languages, he is one of the world's most widely published authors.

Brown is a MacArthur Fellow and the recipient of countless prizes and awards, including the 1987 United Nations Environment Prize, the 1989 World Wide Fund for Nature Gold Medal, and Japan's 1994 Blue Planet Prize for his "exceptional contributions to solving global environmental problems." More recently he was awarded the Presidential Medal of Italy, the Borgström Prize by the Royal Swedish Academy of Agriculture and Forestry, and the Charles A. and Anne Morrow Lindbergh award. He has been appointed to three honorary professorships in China, including one at the Chinese Academy of Sciences. He lives in Washington, D.C.

If you have found this book useful and would like
to share it with others, consider joining our
Plan B Team.

To do so, order five or more copies at our bulk
discount rate at www.earth-policy.org

This book is not the final word. We will continue to
unfold new issues and update the analysis in our
Plan B Updates.
Follow this progress by subscribing to our free,
low-volume e-mail list or RSS feeds at
www.earth-policy.org, and follow us on
Twitter (@EarthPolicy) or on the
Earth Policy Institute Facebook page.

Past Plan B Updates and all of the
Earth Policy Institute's research, including
this book with full endnotes and data sets,
are posted at www.earth-policy.org
for free downloading.

EARTH POLICY INSTITUTE

www.earth-policy.org

.